Potential
Fuel
Effectiveness
in Industry

A Report to the Energy Policy Project of the Ford Foundation

Potential Fuel Effectiveness in Industry

by

Elias P. Gyftopoulos
Massachusetts Institute of Technology

and

Lazaros J. Lazaridis and Thomas F. Widmer
Thermo Electron Corporation

Ballinger Publishing Company ● Cambridge, Mass.
A Subsidiary of J.B. Lippincott Company

Published in the United States of America by Ballinger Publishing Company,
Cambridge, Mass.

First Printing, 1974

Library of Congress Catalog Card Number: 74–16497

International Standard Book Number: 0–88410–309–9–HB
0–88410–311–0–PB

Printed in the United States of America

Library of Congress Cataloging in Publication Data

Gyftopoulos, E   P
          Potential fuel effectiveness in industry.
          (Ford energy series)
          Includes bibliographical references.
          1. Fuel.   I. Lazaridis, Lazaros J., joint author.   II. Widmer, Thomas
F., joint author. III. Title.
TP318.G9   662'.6   74–16497
ISBN 0–88410–309–9–HB
ISBN 0–88410–311–0–PB

Table of Contents

Preface

The Energy Policy Project was initiated by the Ford Foundation in 1971 to explore alternative national energy policies. This book, *Potential Fuel Effectiveness in Industry,* is a technical study of possible methods for conserving fuel. It is one of the series of special studies commissioned by the Project which, I believe, can make a contribution to today's public discussion of energy policies.

At the most, however, each special report deals with only part of the energy puzzle; the Project's own final report, *A Time to Choose: America's Energy Future,* attempts to integrate these parts into a comprehensible whole, setting forth our analysis, conclusions and recommendations for the nation's energy future.

This book, like the others in the series, has been reviewed by scholars and experts in the field not otherwise associated with the Project, in order to be sure that differing points of view were considered. With each book in the series, we offer reviewers the opportunity of having their comments published; none has chosen to do so for this volume.

Potential Fuel Effectiveness in Industry is the authors' report to the Ford Foundation's Energy Policy Project, and neither the Foundation, its Energy Policy Project nor the Project's Advisory Board has assumed the role of endorsing its contents or conclusions. We express our views in our final report, *A Time to Choose.*

S. David Freeman
Director
Energy Policy Project

Potential
Fuel
Effectiveness
in Industry

Chapter One

Introduction

The purpose of this report is to evaluate the potential for more effective use of fuel in U.S. industry.

First, possible saving of fuel with existing technology is estimated in terms of methods that can be applied to industrial processes in general and in terms of methods specific to each of six industries in particular. The six industries — iron and steel, petroleum refining, paper, aluminum, copper and cement — account for about 40% of industrial fuel consumption. Second, the thermodynamic concept of available useful work is employed to determine the ultimate possibilities for saving fuel. This concept allows the calculation of the ideal or minimum fuel requirement for each industrial task and, therefore, the establishment of the maximum saving of fuel which can be expected from entirely new processes that might be developed.

The two most important conclusions of the study are:

(1) With existing technology it is possible to reduce the specific fuel consumption in the six industries studied by one third. If it were realized, such a reduction would approximately offset the fuel needs for the growth of industry that is projected for the remainder of the present decade.

Attention to this opportunity for saving fuel is warranted not only by technological progress with time but also by increased cost and, sometimes, actual shortage of fuel.

(2) Larger saving of fuel may be realized from processes just beyond present horizons of technology. Currently, fuel requirement for a specific industrial practice is often many times the minimum theoretical requirement. Research aimed at more effective industrial processes that consume less fuel per unit of product should, therefore, receive at least as high priority as development of new fuel resources.

Although the economics of some of the technological innovations are evaluated here, neither the time required to implement improved practices

1

with known technology nor the nature of any governmental intervention that might be required to extend them throughout industry has been investigated. We believe, however, that the required technology is so well known that such investigations can and should be undertaken.

The study is based on statistical averages. Of course, the fuel consumption per unit product of each industrial plant may deviate significantly from the corresponding average value for the industry. For this reason, meaningful economic planning for fuel saving should be studied for each major plant in a manner consistent with the relevant production, marketing, environmental and other considerations.

In general, most of the fuel-saving methods discussed in this report require capital investment either for changes in existing plants or for construction of new plants employing the latest technology. A limited number of improvements, however, can be achieved by better operating procedures with existing equipment. Because each industrial plant has different limitations for accommodating changes, we have not classified the proposed fuel-saving methods for each industry into capital demanding and non-capital demanding categories.

The report is organized as follows: Chapter 2 is a summary of the major results of the study. Chapter 3 is devoted to the thermodynamic concept of available useful work and its significance in the evaluation of effectiveness of industrial processes, Chapter 4 discusses general methods for fuel saving, and Chapter 5 presents the reviews of the potential for fuel saving in the six selected industries.

In this study the term fuel, expressed in Btu, refers to the higher heating value of the solid, liquid, and gaseous fuels unless otherwise stated. Electricity is accounted for at the rate of 10,000 British thermal units per kilowatt-hour (Btu per kwhr) regardless of the method of generation. References are listed at the end of each of Chapters 3, 4, and 5.

The Director for this study was Dr. Elias P. Gyftopoulos, Ford Professor of Engineering, MIT, and the Project Manager, Mr. Lazaros J. Lazaridis, Thermo Electron Corporation. Contributors to the study were Dr. Joseph H. Keenan, Professor Emeritus, MIT, Dr. Gregory D. Botsaris, Associate Professor, Chemical Engineering, Tufts University, and Dr. George N. Hatsopoulos, President, Mr. Thomas F. Widmer, Vice President of Engineering, Dr. Sander E. Nydick, Dr. Dean T. Morgan, Mr. Jerry Davis, and Mr. Gabor Miskolczy, all of Thermo Electron.

Chapter Two

Summary

This chapter is a brief summary of the major results emerging from the detailed discussions presented in Chapters 3 to 5. Were the technological means discussed in these chapters to be implemented by U.S. industry they could lead to significant reductions in specific fuel consumption (Btu of fuel per unit of product) in various industrial processes.

Fuels consumed by industry account for about 40% of all fuels consumed in the U.S. They are supplied from all sources such as coal, hydraulic head, uranium, petroleum and natural gas. They are used either in the form of heat for elevating the temperatures of materials and for generating process steam, or in the form of electricity for operating electrical machinery or electrical processes, or in the form of feedstocks for certain industrial products.

In 1968, the amount of fuel consumed by industry was 22.8×10^{15} Btu exclusive of another 2.1×10^{15} Btu equivalent for gas and petroleum used as feedstock materials. It was supplied from the various sources in the amounts shown at the top of Figure 2.1, namely 60% from liquid and gaseous petroleum fuels, domestic and imported, and 40% from coal, hydro, and nuclear fuels which are generally more plentiful and produced from domestic resources. Since distribution of fuel consumption in the U.S. is 76% liquid and gaseous petroleum fuels, and 24% coal, etc, it follows that the industrial sector is relatively more effective than other sectors in its use of plentiful domestic fuels such as coal.

The principal end-uses of fuels in industry can be classified in the four major categories shown at the bottom of Figure 2.1 among which the various fuels are distributed as follows:

1. Direct combustion heating   29.0%
2. Process steam   44.7%
3. Direct electric heating   1.3%
4. Motors, lighting and electrolysis   25.0%

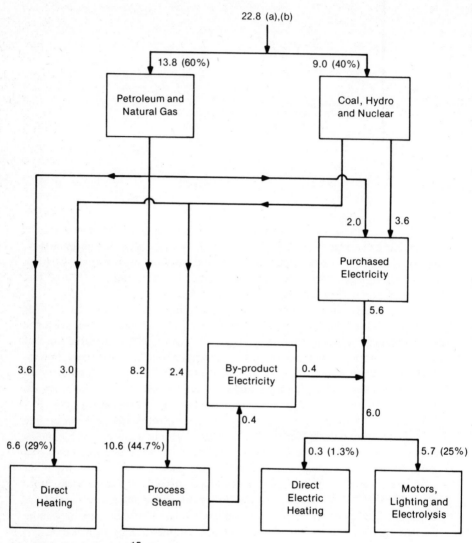

(a) All fuel values in 10^{15} Btu/year

(b) Does not include 2.1 x 10^{15} Btu equivalent fuel
     value for gas and petroleum materials used in
     1968 as feedstock for chemical products.

Figure 2-1.   Sources and end-uses of fuel by U.S. industry in 1968.

Many opportunities exist for the application of existing technology to the management of this enormous fuel flow so as to yield large savings. For example, the bulk of industrial fuel (about 45%) is consumed in raising process steam while almost the entire industrial need for electricity is satisfied by purchased electricity. The combination of raising of process steam with generation of electricity could potentially not only satisfy industrial electrical needs but more importantly result in sale of electricity to the utilities. The potential fuel saving could be as much as 4.0×10^{15} Btu per year at 1968 industrial output levels, or about 30% of all the fuel used by electric utilities in that year (Chapter 4). For a number of reasons which are discussed later, the actual saving would be lower. Nevertheless, the magnitude of the upper limit suggests the existence of a great opportunity for fuel conservation.

Other fuel savings can be achieved through large-scale use of recuperators, regenerators, and low-temperature heat engines for recovery of waste heat from industrial combustion processes (Chapter 4). Recuperators, for example, can cut fuel consumption of radiant-tube heat-treating furnaces by as much as 30%. Again, bottoming-cycle Rankine engines can extract up to 49 kwhrs of electricity per million Btus of enthalpy in the waste heat from Diesels, gas turbines, or furnaces exhausting at $700°F$ or higher.

In addition to the general technologies discussed in Chapter 4, the study has identified specific methods for reduction of fuel consumption in each of the six industries that were studied in relatively more detail (Chapter 5). For 1968, Table 2-1 summarizes the product output in tons per year, specific fuel consumption in Btu per ton of output, and total fuel rate in Btu per year for each of the six industries. Table 2-2, columns 1, 2 and 3 summarize the specific fuel consumption and the specific fuel saving that might be achieved in each of the six industries assuming widespread application of the best technology existing today both in the U.S. and abroad. In addition, Table 2-2, columns 4 to 6 summarize the overall fuel consumption and fuel savings that might be achieved in the six industries if the product outputs were those of 1968. This table illustrates one important conclusion of the study: It is possible, within the framework of present technology, to reduce the specific fuel consumption in the six industries studied by about one third. It is recognized that the projected savings represent an upper limit, and that actual savings which can be realized in practice will be somewhat lower. Nevertheless, the potential is immense and strongly suggests the importance of specific action.

The rate at which the potential fuel savings can be accomplished is, of course, subject to economic considerations including the intricacies of the money market, international relations, institutional arrangements, and social attitudes. Although investigation of this rate is not within the scope of the present study, we believe that the technology is sufficiently well known that such an investigation can and should be undertaken.

The study has also considered the potential for even greater fuel

Table 2-1.   1968 Product Output and Fuel Consumption for Selected U.S. Industries

Industry	Industry Output (tons/yr)	Specific Fuel Consumption (Btu/ton)	Total Fuel Consumption (Btu/yr)	Percentage of Industrial Sector Fuel
Iron and Steel	131×10^6	26.5×10^6	3.47×10^{15}	15.2
Petroleum Refining	590×10^6	4.4×10^6	2.6×10^{15}	11.4
Paper and Paperboard	50×10^6	*24.5×10^6	1.22×10^{15}	5.4
Aluminum (primary & scrap)	4.07×10^6	155×10^6	0.63×10^{15}	2.8
Copper	3.1×10^6	25.8×10^6	0.08×10^{15}	0.4
Cement	72×10^6	7.9×10^6	0.57×10^{15}	2.5
		Total	8.6×10^{15}	38%

*Does not include heating value of waste products (bark and spent pulp liquor). If waste products were included, specific fuel consumption would be 39×10^6 Btu/ton of paper.

Table 2-2.   1968 and Improved Fuel Consumption for Selected U.S. Industries

Industry	Specific Fuel Consumption (10^6 Btu/ton)		Percentage Improvement in Specific Fuel Consumption Over 1968 Practices	Industry Output (1968) (10^6 tons/yr)	Total Fuel Consumption (10^{15} Btu/yr)	
	With 1968 Practices	With Potential Process Improvements (technology existing in 1973)			With 1968 Practices	With Potential Improvements (1973 technology)
Iron and Steel	26.5	17.2	36%	131	3.47	2.26
Petroleum Refining	4.4	3.3	25%	590	2.6	1.95
Paper and Paperboard	39.0*	23.8*	39%	50	1.95	1.2
Aluminum (primary and scrap)	155	106.0	32%	4.07	0.63	0.43
Copper	25.8	18.1	33%	3.1	0.08	0.05
Cement	7.9	4.5	43%	72.0	0.57	0.32
TOTALS					9.3	6.2
FUEL SAVING						33%

*Includes waste products consumed as fuel by paper industry.

savings in later years through development of radically new processes which more nearly approach the theoretical limits of minimum specific fuel consumption. Table 2-3 lists values of specific fuel consumption (Btu/ton) for several industries based upon 1968 practices, suggested practices using currently demonstrated technology, and the theoretical limits computed by means of the concept of available useful work (Chapters 3 and 5). The large margins that exist between current practices and minimum theoretical requirements indicate the potential which is available for major long-term reductions in fuel consumption through basic process modifications. Research aimed at more effective industrial processes that consume less fuel per unit product should, therefore, receive at least as high a priority as development of new energy resources.

Table 2-3.   Comparison of Specific Fuel Consumption of Known Processes With Theoretical Minimum for Selected U.S. Industries

	1968 Specific Fuel Consumption (Btu/ton)	Potential Specific Fuel Consumption Using Technology Existing in 1973 (Btu/ton)	Theoretical Minimum Specific Fuel Consumption Based Upon Thermodynamic Availability Analysis (Btu/ton)
Iron and Steel	26.5×10^6	17.2×10^6	6.0×10^6
Petroleum Refining	4.4×10^6	3.3×10^6	0.4×10^6
Paper	*39.0×10^6	*23.8×10^6	[†]Greater than -0.2×10^6 Smaller than $+0.1 \times 10^6$
Primary Aluminum Production**	190×10^6	152×10^6	25.2×10^6
Cement	7.9×10^6	4.7×10^6	0.8×10^6

*Includes 14.5×10^6 Btu/ton of paper produced from waste products consumed as fuel by paper industry.
**Does not include effect of scrap recycling.
[†]Negative value means that no fuel is required.

Chapter Three

Thermodynamic Availability Analysis

This chapter discusses the thermodynamic concept of available useful work which can be used both in evaluating the effectiveness of the use of fuel and in calculating the minimum cost for any process. Readers unfamiliar with the concepts of thermodynamics might consult the article on "Thermodynamics" in the 1974 Edition of the Encyclopedia Britannica.

3.1 AVAILABLE USEFUL WORK—DEFINITION

The principles of thermodynamics indicate that the measure of effectiveness with which fuel is used in various industrial processes requires consideration of properties other than energy alone. For example, every engineer knows that a Btu of enthalpy in the circulating coolant water of a powerplant has less value than a Btu of enthalpy in the high temperature steam main. Similarly, it is obvious that a cold but fully charged storage battery is more useful than a discharged battery having the same total energy by virtue of its being hot.

These examples illustrate the need for a yardstick other than energy to assess the minimum fuel needs of any particular process. The laws of thermodynamics indicate that the relevant quantity is a property called available useful work which, in turn, is uniquely related to another property called entropy. Associated with every system in any given state is an amount of work which is the minimum needed to create the system, in that state, out of materials in the atmosphere or in an ocean in equilibrium with the atmosphere. This minimum amount of work is also equal to the maximum work that can be done by the system starting from the given state and ending in a state of mutual stable equilibrium with the atmosphere. Each of these works is called available useful work.

Values of available useful work are, of course, associated with the fuels we use. If a fuel such as the molecular species CH_2 were to be formed in the best way possible, namely reversibly, from carbon dioxide and water in the atmosphere, the work required would be the available useful work of the fuel. This work could be recovered completely for use on other systems if the fuel were combined with oxygen from the air in the best way possible, namely in a reversible process, which restores the carbon dioxide and water as constituents of the gaseous mixture we call air. It is appropriate therefore to give thought to the minimum magnitude of the oxidation process that will permit any specified task to be performed—that is, the minimum fuel requirement. This minimum would be attained if the oxidation process and all subsequent operations were to be executed reversibly within the terrestrial (air and water) environment. It can be shown that the minimum requirement for all the tasks that are performed in the U.S. economy would consume a small fraction of the fuel now being used.

The available useful work or task-performing value of mixtures of fuel and air is dependent upon the temperature and other properties of the environment (ambient air and water) to which or from which the fuel-air system may deliver or receive energy. In a reversible process the available useful work is conserved; that is, when fuel is oxidized in order to perform a specific task on a specific material, the available useful work is merely transferred from the fuel-air system to the material.

The available useful work Φ of a system and atmosphere can be shown to be given by the equation

$$\Phi = E + p_o V - T_o S - \sum_{i}^{n} \mu_{io} n_i$$

where E denotes energy, V volume, and S entropy of the system, n_i for $i = 1, 2, \ldots n$ number of moles of molecular component i in the system, p_o the pressure of the atmosphere, T_o the temperature of the atmosphere, and μ_{io} the total potential of component i in the atmosphere or in mutual stable equilibrium with it. The quantity Φ may be evaluated for any system in any state whether stable equilibrium, nonstable equilibrium, or nonequilibrium. In particular, its value is zero when the system and the atmosphere are in mutual stable equilibrium. That is, Φ is zero when the system is in a stable equilibrium state such that its temperature is T_o, its pressure is p_o, and its total potentials are μ_{io} for $i = 1, 2, \ldots n$. In general, the available useful work is different from another property called Gibbs free energy.

For given energy, volume, and composition of the system, Φ decreases as entropy of the system increases. Because a deficiency in available useful work corresponds to a surplus of entropy, any increase in entropy during a process which is not the best possible, namely during an irreversible process, decreases the available useful work. The value of Φ may exceed the energy of the system although it is always smaller than the energy of the system and the atmosphere taken together.

The available useful work Φ was devised by Gibbs[1] in 1875. In an earlier paper Gibbs[2] had introduced the function

$$\phi = E + p_o V - T_o S$$

for application to a system the constituent substance of which did not enter into or mix with the atmosphere. The maximum possible decrease in the value of this function as the system (body) proceeded toward pressure and temperature equilibrium with the atmosphere (medium) he called the "available energy of the body and medium."

A closely related property

$$H - T_o S,$$

which is particularly useful for calculating the available work in steady flow processes, was used by Darrieus[3] in 1930 for processes in turbines and by Keenan[4] in 1932 for analysis of steam powerplants and for cost accounting when both process steam and power are produced. It appeared in an engineering textbook by Bosnjakovic[5] in 1935 and more recently in European literature[6] where it has been given the name "exergy."

When applied to a hydrocarbon fuel, the quantity Φ is the minimum useful work required to form the fuel in a given state from the water and carbon dioxide in the atmosphere. Since this minimum will be the useful work of a reversible process, the quantity Φ is also the maximum useful work which could be obtained by oxidation of the fuel and return of the products to the atmosphere. Moreover, any change in state of the fuel-air-atmosphere system will produce an amount of useful work less than, or in a reversible change equal to, the corresponding decrease in Φ.

The measure of effectiveness of an industrial process is the increase in available useful work between raw materials entering and industrial products leaving the process, divided by the available useful work of the fuel consumed. If the products leave the process at high temperature, as in a blast furnace, the process would not then be charged with available useful work in the products by virtue of the elevated temperature. If this work is lost, the deficit is not in the process but in the means by which the products are cooled to environmental temperature. Various examples illustrating the concept of available useful work in a number of physical and chemical processes are discussed in the following sections. In particular, the examples given in Sections 3-2 and 3-3 illustrate the importance of the concept in evaluations of effectiveness of fuel utilization and technico-economic optimization studies.

3.2 FUEL OXIDATION AND HEAT TRANSFER PROCESSES

The curves of Figure 3-2-1 are calculated for one pound-mole of a liquid hydrocarbon fuel which may be described as CH_2 and which has a lower

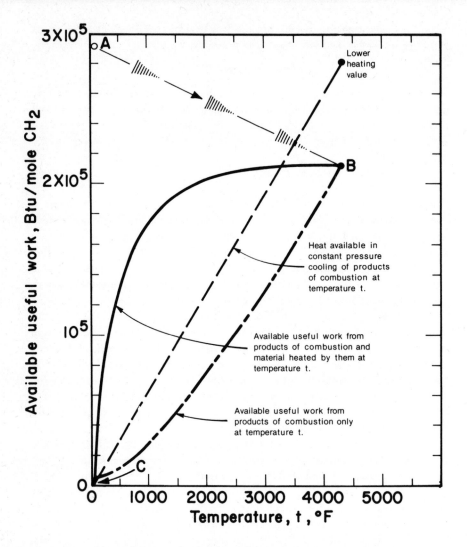

Figure 3-2-1.   Available useful work from hydrocarbon oxidation process.

"heating value" of 280,000 Btu (higher heating value = 298,000 Btu). This heating value is the decrease in enthalpy when one pound-mole of fuel burns in air to form carbon dioxide, water and nitrogen:

$$CH_2 + \frac{3}{2}O_2 + 5.65N_2 \rightarrow CO_2 + H_2O + 5.65N_2$$

The available useful work for the reactants can be shown to be about 291,000 Btu per pound-mole, namely about 3% greater than the lower heating value. In order to obtain this work the following process, as one among many possibilities, could be used: (1) the oxidation is carried out in a reversible fuel cell, at some temperature T_A, which delivers electrical work to the surroundings; (2) the products of the fuel-cell process are cooled from T_A to the temperature of the environment T_0 as they provide heat to Carnot engines which produce further work; (3) each of the products CO_2, H_2O and N_2 is separated from the mixture reversibly by means of a semipermeable membrane and expanded reversibly and isothermally at T_0 in an engine cylinder until it attains a pressure equal to the partial pressure of that constituent in the atmosphere; (4) each molecular species is introduced reversibly into the atmosphere through a semipermeable membrane. The term semipermeable membrane refers to a device which is impermeable to all molecular species except one. The final state after step (4) above corresponds to zero available useful work at atmospheric temperature ($55°F$). It is the base state C in Figure 3-2-1.

Such a fuel-air process would produce the maximum possible work, the available useful work. Moreover, the work so obtained could be used in an inverse process to create the original quantity of fuel from the carbon dioxide and water vapor in the air. Any oxidation process which produced less work than the available useful work would be irreversible, and the loss in available useful work would be a measure of the irreversibility of the process.

Because fuel cells for efficient oxidation of a hydrocarbon fuel are not currently obtainable, although they are in various stages of development, fuels are almost always burned in a combustion chamber without production of electrical current. For CH_2 and the reaction cited above, the temperature at the end of the combustion process is about $4300°F$, the adabatic flame temperature for the stoichiometric mixture. Being irreversible, the combustion process, suggested in Figure 3-2-1 by the broken line AB, results in an increase in entropy and a loss of available useful work, namely in a loss of ability to do work without a corresponding loss in energy.

Here the loss is 80,000 Btu per pound-mole of our assumed fuel or ⁿbout 27% of the original value of 291,000 Btu. The remaining available useful vork is 211,000 Btu, which is the maximum amount of work which could be obtained, for example, by transferring heat to Carnot engines and by expansion of the product species to the limit imposed by the environment at C.

The combustion process is a constant-enthalpy process; that is, the capacity for solely transferring heat in steady flow to surrounding systems remains unaltered by the combustion process. By virtue of the irreversibility and the associated increase in entropy in the adiabatic combustion process, available useful work has been lost at constant enthalpy. An analogous process is the flow of a perfect gas through a throttle valve from high pressure to low. Enthalpy remains constant while entropy increases. In the throttling process available useful work is lost while temperature remains constant, whereas in the combustion process available useful work is lost while the temperature rises. Both are adiabatic processes in which entropy increases because of irreversibility.

Beginning with state B in Figure 3-2-1, the available useful work can be altered in a number of ways. For example, energy from the combustion products may be transferred in the form of heat to any material at a temperature t less than 4300°F, and the transferred energy may be used reversibly to produce work. Because the temperature difference between the combustion products and the material is finite, the transfer process is irreversible and, therefore, the available useful work decreases. The solid curve in Figure 3-2-1 shows the available useful work in the products plus that in a material at temperature t°F which has cooled the products to t°F without itself changing temperature. For this purpose the material is assumed to be of infinite heat capacity because otherwise some heat would be transferred to the material at temperatures less than t°F and the loss in Φ would be correspondingly greater; that is, the solid curve shows the maximum available useful work after cooling the products to t°F.

The dashed curve shows the heat available from products of combustion in simple steady-flow cooling from temperature t to the temperature of the atmosphere (55°F). It is, indeed the variation of enthalpy with temperature. The dot-and-dash curve shows the corresponding available useful work of the products of combustion at temperature t. It reaches atmospheric temperature at a small positive value of the ordinate corresponding to the work obtainable upon reversibly mixing the products with the constituent gases of the atmosphere. The difference in ordinate between the solid curve and the dot-and-dash curve at any value of t is the available useful work from infinite-heat-capacity material which has cooled the products from 4300°F to t°F.

It is evident from the solid curve that as the temperature of the heat-receiving material is lowered below 2000°F the loss in Φ increases rapidly with decrease in temperature. At a temperature of about 600°F, the value of Φ is about 48% of that for the fuel initially, or 140,000 BTU per mole of fuel.

A typical average temperature of the heat-receiving water-steam working fluid in a central steam powerplant is 600°F. Accordingly, about 25% of the available useful work of the fuel is lost in the irreversible process of transferring energy from the products of combustion to the working fluid across a finite temperature difference. Magneto-gas-dynamic and thermionic devices have been proposed to bridge these temperature differences and to salvage losses in Φ by reducing irreversibilities.

In an industrial plant, on the other hand, the products of combustion might be used to make process steam at, say, 270°F. The loss of available useful work is then 291,000 minus 90,000 or about 69% of that in the fuel. The difference between 69 and 52% represents the fraction lost because a steam powerplant was not interposed between the products of combustion and the process steam. It can be shown by a simple calculation that for the same amount of process steam an interposed steam powerplant may produce 60,000 Btu of electrical work for an additional 0.23 moles of fuel consumed. Thus, 60,000 Btu of electrical work would be obtained at the expense of 0.23 x 291,000 or 67,000 Btu of available useful work. A central station powerplant would consume about 0.55 moles of CH_2 (as compared with 0.23) and 161,000 Btu of available useful work (as compared with 67,000) to produce the same amount of electrical work.

The preceding paragraphs discuss an example of topping a heating process with a power-producing process in order to reduce the loss of available useful work. Many industrial heating processes, on the other hand, require such high temperatures that a topping process would require either unobtainable or prohibitively expensive materials. Moreover, the saving to be realized per thousand degrees of temperature interval, as shown by the upper portion of the solid curve in Figure 3-2-1, is small.

In these high-temperature processes, such as the manufacture of steel or cement, emphasis should be placed on salvaging available useful work from the material in process and from the products of combustion leaving the heating process. For example, the billets leaving a heating furnace contain 15 to 50% as much potential to supply heat to other processes as the fuel originally consumed. A similar range of potential may be found in the products of combustion leaving the heating process. The opportunities for reclaiming available useful work through heat transfer to power-producing or lower-temperature heating processes are evident.

3.3 OXYGEN SEPARATION PROCESS

The available useful work Φ of oxygen and residual gases—mostly nitrogen—separated from one pound mole of air at a pressure of one standard atmosphere and a temperature of 80°F is 557 Btu. This is, of course, the minimum work required within an environment at one atmosphere and 80°F to produce from that atmosphere the two separate gases at the environmental pressure and temperature.

The amount of work used in a practical oxygen separation process is much greater than this minimum. It has been analyzed by Manson Benedict[7] and his results are given below. Figure 3-3-1 illustrates the process flow for one type of oxygen production cycle considered by Benedict. The process quantities refer to separation of 10,000 moles of air per hour into a waste fraction containing 99% nitrogen and a product fraction containing 90% oxygen. The plant produces 380 tons of oxygen per day with a recovery of 96%.

Air at 77°F is compressed by a blower to a pressure of 20.7 psi, which is sufficient to force the air through the exchangers and towers of the

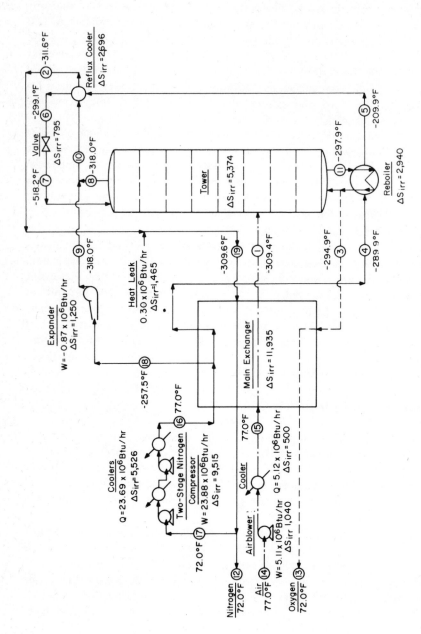

Figure 3-3-1.   Practical oxygen separation process.

separation plant. In the main exchanger, the air is cooled by means of the outgoing product oxygen and nitrogen, and emerges at its dew point at $-309.4°F$. In the tower, the air is fractionated into the nitrogen overhead and oxygen bottom streams. These streams are returned through the main exchanger and discharged from the plant at $72°F$. The nitrogen is also used to subcool tower reflux and is finally discharged at atmospheric pressure. The oxygen, which emerges from the tower at a somewhat higher pressure than the nitrogen, is discharged from the plant at 16.7 psi.

To reboil the tower, to provide reflux and to satisfy the refrigeration requirements of the plant, an auxiliary nitrogen stream amounting to 90% of the air fed to the plant is compressed in a two-stage compressor to 77 psi and is also cooled to its dew point of $-289.9°F$ in the main exchanger. This nitrogen is liquefied in the reboiler, where its latent heat reboils the tower, is subcooled against outgoing product nitrogen in the reflux cooler and finally is flashed through a valve into the top of the tower where it provides reflux at $-318.2°F$. In the tower, the nitrogen reflux stream is vaporized into the product nitrogen and flows with it through the nitrogen pass of the main exchanger where it gives up its heat to the incoming compressed nitrogen. To compensate for the heat leak to the plant and the enthalpy difference between the outgoing nitrogen and oxygen streams and incoming air, a portion of the compressed nitrogen is made to do work in the expander in order to lower its temperatures from $-257.5°F$ and $-318.1°F$.

In setting up the conditions for this process, Benedict assumed the following: (a) pressure drops of 2 psi through the main exhanger and 1 psi through the other exchangers and from top to bottom of the tower; (b) a temperature difference of $5°F$ for each of the gas-to-gas exchangers; (c) a net heat upflow in the tower equal to 3.5% greater than the minimum flow needed to carry out the separation with an infinite number of plates; (d) a heat leak of 30 Btu/mole of incoming air or 300,000 Btu/hr; and (e) efficiencies for the compressors, the blower and the expander equal to 75%. Thus, he finds that a total of 29×10^6 Btu/hr of work is expended in the nitrogen compressors and the blower, and 0.87×10^6 Btu/hr of work is recovered in the expander. The net work input is 28×10^6 Btu/hr or 10.8 kwhr per 1000 cubic feet of oxygen. The theoretical work of carrying out this separation, evaluated from the change in enthalpy and entropy of the feed and products, is only 4.6 million Btu/hr or 1.7 kwhr per 1000 cubic feet of oxygen.

Next, Benedict considers the process inefficiencies which are responsible for the increased work expenditure. This is done by evaluating the entropy production of each element of the process, namely by evaluating the irreversibilities. The largest portions of this entropy production, shown in Figure 3-3-1, are the main exchanger, the nitrogen compressors, the nitrogen coolers and the tower.

To improve the effectiveness of the process, it is necessary to determine the extent to which the entropy production is characteristic of the process and the extent to which it is a consequence of equipment inefficiencies which could be reduced by increased capital expenditure for either larger or more effective equipment. To do this, Benedict considers an idealized process design for a plant in which the exchangers are so large that the pressure drop is zero and the temperature difference is the minimum consistent with process heat balance. In addition, he assumes that the heat leak may be made negligible by elaborate insulation, that only the minimum heat needed to reboil the tower will be used, and that the nitrogen compressor may be made 100% efficient and isothermal.

Figure 3-3-2 shows the flow sheet for such an idealized process. The work input is reduced from 28×10^6 Btu/hr to 7.7×10^6 Btu/hr, and the irreversible entropy production has been made practically negligible in all pieces of equipment except the tower. The entropy production here is still high because of the fact that fractionation occurs away from equilibrium conditions at all points of the tower except at the feed point.

Table 3-3-1 compares the work input and rate of entropy production in the practical process with corresponding quantities in the ideal process. This brings out the fact that the single place where the greatest reduction in power can be effected is in the main exchanger, and also shows the relative importance of the individual pieces of equipment as contributors of the total work input. It also shows that the total work input is the sum of the theoretical work and the work equivalent to the irreversible rate of entropy production within the process.

To illustrate how the entropy production may be used to select approximate optimum design conditions for individual units of process equipment, Benedict analyzed the conditions selected for the main exchanger from the standpoint of their contribution to the cost of the exchanger, the cost of the compressors, and the cost of the power required to run the plant for five years. His results are given in Figure 3-3-3. The figure illustrates the cost balance on this exchanger. The pressure drop and the warm end temperature difference for which the exchanger is designed contribute to the cost of the plant in two principal ways. First, the lower the pressure drop and temperature difference, the larger the exchanger and the greater its cost. Second, the lower the pressure drop and warm end temperature difference, the lower the irreversible rate of entropy production in the main exchanger, the smaller the compressors needed for the plant and the lower the cost of power supplied to the plant. The equation given in the figure indicates how the exchanger surface (A) will vary with pressure drop and temperature difference. The assumed cost for exchanger surface is $1 per square foot.

Benedict translated the effect of pressure drop and temperature difference in increasing the rate of entropy production from the ideal case to the practical case to the equation given in Figure 3-3-3. He assumed that the initial cost of the compressors is increased by $50 for every kilowatt increase in power

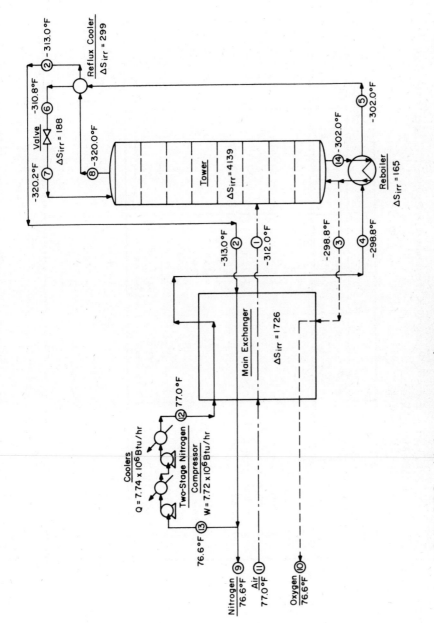

Figure 3-3-2.   Ideal oxygen separation process.

Table 3-3-1.   Entropy Production Breakdown in Oxygen Processes. Feed Rate: 10,000 Moles Air/hr

Process	Ideal		Practical	
	Work Input, 10^6 Btu/hr	ΔS_{irr}, Btu/°R hr	Work Input, 10^6 Btu/hr	ΔS_{irr}, Btu/°R hr
Air Blower	7.72	--	5.11	1,840
Air Cooler		--		500
N_2 Compressor		--	23.88	9,515
N_2 Coolers		--		5,526
Main Exchanger		1,726		11,935
Expander		--	-0.87	1,250
Tower		4,139		5,374
Reboiler		165		2,948
Reflux Cooler		299		2,696
Valve		188		795
Heat Leak		--		1,465
Total	7.72	6,517	28.12	43,844
$T_o \Delta S_{irr}$, 10^6 Btu/hr		3.50		23.54
Theoretical Work = $\Delta H - T_o \Delta S$		4.21		4.61
Predicted Work, 10^6 Btu/hr		7.71		28.15

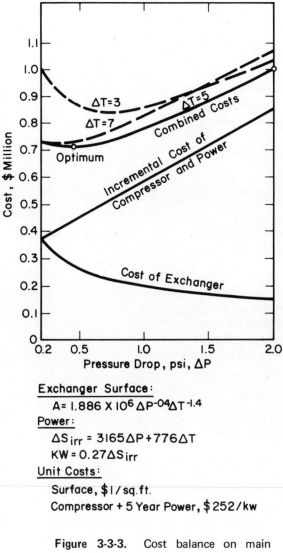

Figure 3-3-3. Cost balance on main exchanger.

input and that power is to be charged at the rate of 0.6¢/kwhr for a period of five years. Thus he finds that the total cost of power is $252 per kilwatt or $68 per unit rate of entropy production.

The solid curves in Figure 3-3-3 show the effect of pressure drop upon cost for a warm end temperature difference of 5°F. The opposing tendencies of entropy production and exchange surface are indicated. Total cost is minimized at a pressure drop of about 0.5 psi rather than the 2 psi used in the practical process. Similar curves for the total cost at temperature differences of 3° and 7°F indicate that a warm end temperature difference of 5°F is close to the optimum.

Thus, by consideration of the rate of entropy production in the main exchanger, the optimum pressure drop and temperature difference across this exchanger may be estimated without the necessity for a complete system design at each combination of conditions.

REFERENCES

1. J. W. Gibbs, Collected Works, Longmans and Yale Univ. Press, Vol. 1, p. 77, Equation (54).

2. Ibid., p. 58.

3. G. Darrieus, Revue Generale de l'Electricite, June 21, 1930, Vol. 27, 963-968; also Engineering, Sept. 5, 1930, 283-285.

4. J. H. Keenan, Mechanical Engineering, March 1932, Vol. 54, 195-204.

5. Fr. Bosnjakovic, Technische Thermodynamik, Steinkopff, 1935, Vol. 1, p. 138; Vol. 2, p. 2.

6. Brennstoff-Warme-Kraft, Fachheft Exergie, Nov. 5, 1961, No. 11.

7. Manson Benedict, Massachusetts Institute of Technology. Unpublished lecture given at MIT in April 1949.

Chapter Four

General Methods
for Fuel Saving

This chapter presents a number of proven fuel-saving techniques that can be applied to industrial processes in general.

4.1 STEAM RAISING

4.1.1 Combined Steam and Electricity Generation

Wherever process steam is required, an opportunity exists to produce electrical work at small cost in fuel consumed. For example, if process steam at 200 psi or $382°F$ is generated by burning a hydrocarbon fuel, $(CH_2)_n$, according to Figure 3-2-1, the available useful work lost is $291,000 - 115,000$ or $176,000$ Btu per mole of CH_2. Much of this loss may be prevented by burning the fuel in a gas turbine and using the turbine exhaust to generate steam (Figure 4-1-1a) by generating steam at a pressure higher than 200 psi and expanding the steam in a steam turbine to 200 psi at which pressure it is exhausted to process (Figure 4-1-1b), or by a combination of these two (Fig. 4-1-1c).

In order to show the amount of electrical power which can be generated, and the low cost in fuel consumed for the power produced by these means, several examples have been calculated and the results displayed in Table 4-1-1. These calculations are based upon the following assumptions: the steam-power cycle efficiency† is about 80% of the corresponding Rankine-cycle efficiency; the average temperature of the heat input to the steam cycle is $600°F$; the gas turbine efficiency is 22.5% based on the lower heating value of the fuel; and electric generator efficiency is 90%.

The numbers in Table 4-1-1 give the power produced in kilowatts for a rate of heat to process steam of 10^6 Btu per hour or about 7 gallons per hour

---

†Efficiency of a heat engine is defined as the ratio of the work output of the engine to the heat input of the engine.

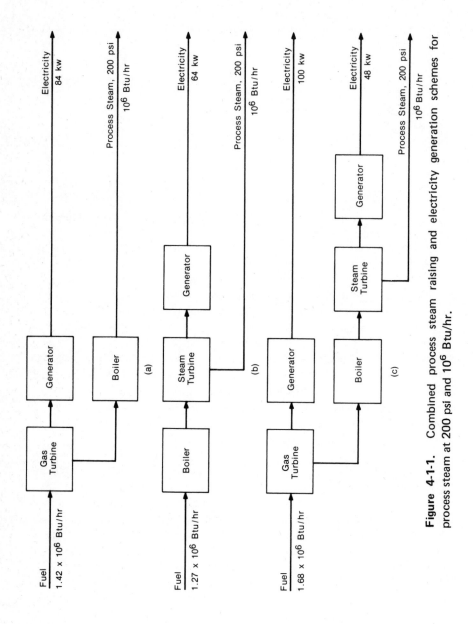

Figure 4-1-1. Combined process steam raising and electricity generation schemes for process steam at 200 psi and 10^6 Btu/hr.

Table 4-1-1. Kilowatts of Power for 10^6 Btu/hr of Heat to Process

Process Steam Pressure, psi	Steam-Turbine Power, kw		Gas-Turbine Power, kw		Total Combined Gas and Steam-Turbine System Power, kw
	Alone	Fed From Exhaust of Gas Turbine	Alone	Followed by a Steam Turbine	
50	77		84		
200	49	48	84	100*	148
400	34		84		

*The power of the gas turbine is increased from 84 to 100 kw because some of the available useful work of the fuel necessary for the steam turbine is consumed in the gas turbine.

of petroleum fuel. The power produced, if considered as a by-product of the process heat, should be charged with the fuel consumption over and above that required when process steam is produced directly without intervening power-producing machinery. On this basis, the fuel cost for each of the cases shown in Table 4-1-1 is about 4230 Btu of available useful work in fuel for each kwhr of electrical work. The corresponding figure in terms of higher heating value of the fuel is 4325 Btu per kwhr. These figures translate into an effectiveness of 0.8 and an efficiency of 0.79. The corresponding figures for the best central-station powerplants are 0.39 and 0.38; that is, power is produced at half the fuel cost of the best central-station powerplants.

For an efficiency (based on higher heating value) of 0.79 the fuel cost per kwhr of electrical work may be calculated for any selected price of fuel. For a price of 14 cents per gallon of fuel oil, corresponding to about $0.94 per million Btu of higher heating value, the fuel cost proves to be 0.41 cents per kilowatt hour of electricity. To this, of course, must be added the capital charges corresponding to the increase in capital charges incident to the by-product power. In Table 4-1-2 such costs are tabulated for a low-pressure boiler cost of $10.00 per kilowatt of heat rate into the generated steam, high-pressure boiler cost of $20.00 per kilowatt, a steam-turbine cost of $50.00 per kilowatt of electrical power, and a gas-turbine cost of $100.00 per kilowatt. For 400 psi process, the boiler cost per unit of heat was assumed to be the same with or without by-product power. Capital charges were figured at 20% per year and distributed over full power output for one-half the hours of the year.

In Table 4-1-2 the capital charges range from 0.33 to 0.60¢/kwhr and are much the same for steam turbines and gas turbines. The total of fuel and capital charges ranges from 0.74 to 1.01¢/kwhr. It would seem reasonable to put

Table 4-1-2.   Capital and Fuel Charges Against Electrical Power as a By-Product of Process Heat†

Process Pressure, psi	50	200	400	200	200
Power Unit	Steam Turb.	Steam Turb.	Steam Turb.	Gas Turb.	Steam & Gas Turb.
Heating value to boiler x 10^{-6}, Btu/hr	1. 37	1. 27	1. 20	1. 0††	1. 23†††
By-product power, kw	77	49	34	84	148
Boiler cost with power	7560	6960	6600	2930	7180
Boiler cost without power	2930	2930	5860	2930	2930
Boiler increm. cost, $	4630	4030	740	0	4250
Steam-Turbine cost	3850	2450	1700	-	2400
Gas-Turbine cost	-	-	-	8400	10, 000
Total increm. capital	8480	6480	2440	8400	16, 650
Capital charge, ¢/kwhr	0.50	0. 60	0. 33	0. 46	0. 51
Fuel charge, ¢/kwhr	0. 41	0. 41	0. 41	0. 41	0. 41
Capital and fuel charge, ¢/kwhr	0.91	1. 01	0. 74	0. 87	0. 92

†All values given are for 10^6 Btu per hour delivered to the process steam; fuel input rates are based upon higher heating values.

††Heating value to gas turbine 1.42 x 10^6 Btu per hour.

†††Heating value to gas turbine 1.68 x 10^6 Btu per hour.

the cost in practice at approximately 1¢/kwhr, which may well be a profitable figure at pre-Arab embargo fuel costs. If fuel costs were to double, this figure would increase by less than 50%, whereas costs in a central-station plant would doubtless increase by a larger percentage because 1 kwhr generated by the combined plant requires about 4300 Btu whereas that generated by a central-station plant requires 10,000 Btu.

The cost figures in Table 4-1-2 include no incremental labor charge. If any, it should probably be added as a charge per hour of power generation.

Thus, for an incremental charge of $1.00 per hour applied to the operating hours, the labor cost would be 100/P cents per kilowatt hour when P kilowatts are generated. For values of P in excess of 1000 kw, the labor cost need not be prohibitive.

It is probable that a lower grade of fuel could be used for steam power than for gas-turbine power. For example, coal could be used for the former but not for the latter. For lower pressures of process steam, therefore, steam turbines could well be preferable to gas turbines. At higher pressures, power from steam turbines would in some instances be too small for good economy; for example, at 400 psi the power generated is less than half that at 50 psi. The gas turbine, on the other hand, will generate the same power for process steam at both low and high pressures.

The above analysis indicates that an average of about 70 kw of electric power can be obtained for every million Btu per hour of steam required by industrial processes. In terms of central-station fuel to supply the same amount of power, the by-product power displaces $(70 \times 3412)/0.34$ or about 700,000 Btu/hr of fuel heating value. In 1968, U.S. industry consumed about 10.2×10^{15} Btu for process steam and only about 10^{11} kwhr of by-product electricity was generated (a rate of about 10 kw for each million Btu per hour). In principle, this steam flow could have given 7×10^{11} kwhr (70 kw per million Btu per hour), or about 53% of total U.S. electrical production in that year. The net fuel savings (fuel saved by utilities less the incremental fuel consumed by industry for electricity generation) would have been 4.0×10^{15} Btu or 30% of all fuel used by the electric utilities.

Obviously, many industrial operations requiring process steam are too small in scale to permit economical generation of by-product electricity so that not all of the above potential can be realized. Nevertheless, an opportunity exists for an enormous fuel saving if such practices are employed on a broader scale. This will require the development of markets for by-product power through utility networks, since many industries will have to become net producers of electricity. In other words, the expanded production of by-product power will not be aimed at the electrical needs of industry, but at the national problem of fuel conservation. For this reason, accommodations not now existing between industry and public utilities will be called for.

4.1.2 Use of Solar Energy for Raising Process Steam

Relatively simple flat-plate solar collectors with special coatings are capable of heating fluids to temperatures of $180-250°F$, depending upon the available incident solar flux. Such a collector can generate saturated steam at approximately $180°F$ and 7.5 psi. This steam can then be compressed and superheated by means of an efficient engine-compressor to yield three times as much process steam per unit of fuel consumption as that generated by direct firing.

The cycle shown in Figure 4-1-2, which has a Diesel engine as the prime mover, will provide 1.93 pounds of steam at 668°F and 65 psi for each 1000 Btu of heating value in fuel to the engine. Engine exhaust, engine cooling, and compressor losses all contribute to the enthalpy of the process steam. Direct firing of a boiler with an efficiency† of 85% will generate only 0.635 pounds of process steam per 1000 Btu of fuel at the above conditions. Where steam of higher pressure is required, the performance advantage of the Diesel-solar system is reduced. For example, the three-fold advantage over a directly-fired system at 65 psi is reduced to a two-fold advantage at 215 psi.

Although sunshine is intermittent and unreliable, terrestrial solar energy systems can employ either thermal storage, a backup conventional system, or both. Because the capital cost of a conventional boiler is relatively low compared to fuel costs, it is possible to avoid all thermal storage by including 100% capacity in conventional stand-by equipment. Diesel engines can operate on low-grade residual fuels. Operation on "Bunker C" is common for marine Diesels which have long life, good reliability, and low maintenance cost. Net efficiencies from fuel to shaft power are commonly 38% to 40%.

An estimate of the first cost and yearly operation cost, based on a single eight-hour daily shift, for conventional and Diesel-solar without thermal storage systems is given below. Peak solar flux for U.S. latitudes is approximately 1 kw/m². An average input at 180°F for eight hours is taken to be 0.4 kw/m² or 127 Btu/hr-ft². First costs are assumed to be as follows.

Conventional Burner/Boiler	$ 10/kwth
Waste Heat Recovery Boiler	$ 10/kwth
Diesel Engine	$100/kwe
Compressor	$ 50/kwe
Solar Collectors	$ 5/ft²
Fuel	80¢/10⁶ Btu
Capital Charges	15%

Table 4-1-3 gives corresponding cost figures for a fuel-consumption rate in the Diesel engine of 10^6 Btu of heating value per hour.

When the listed prices are used, the calculated annual cost for the Diesel-solar system is about twice as much per unit of steam generated as that for the conventional system. Figure 4-1-3 shows the effect of varying fuel cost and solar-collector first cost. For collector costs at $2.50 per square foot, the breakeven point for economic Diesel-solar steam raising should occur at a fuel cost of about $1.30 per 10^6 Btu. Operating, labor and maintenance costs have not been included in this analysis. A typical Diesel-generator maintenance cost of 0.15/kwhr would add about $500 to the Diesel-solar annual cost, if other

---

†Efficiency of a boiler is defined as the ratio of the enthalpy of the process fluid over the enthalpy of the fuel consumed.

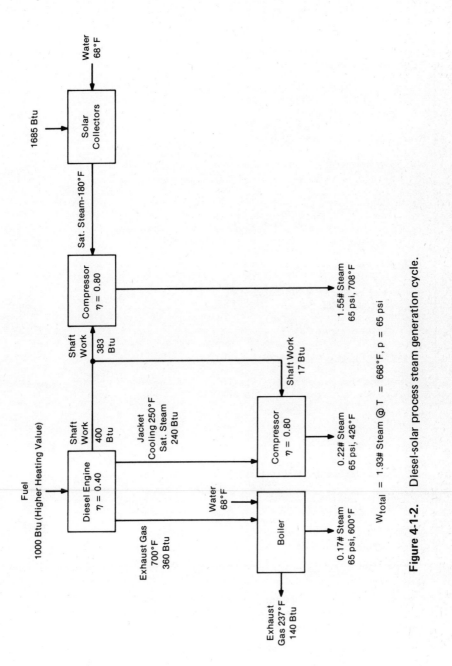

$W_{total} = 1.93\#$ Steam @ T $= 668°F$, p $= 65$ psi

Figure 4-1-2. Diesel-solar process steam generation cycle.

Table 4-1-3.   Cost for Diesel-Solar and Conventional Systems for Process Steam Generation

Solar Collector Cost		
Solar collector flux requirement	1.685×10^6 Btu/hr	
Area ($1.685 \times 10^6/127$)	1.33×10^4 ft^2	
Cost ($5 \times 1.33 \times 10^4$)		$ 66,500
Diesel and Compressor Cost		
Shaft power (0.4×10^6 Btu/hr)	117 kw	
Cost (150×117)		17,700
Waste Heat Boiler		
Heat recovery (0.36×10^6 Btu/hr)	105 kwth	
Cost (10×105)		1,050
100% Standby Burner-Boiler		
Power rating (3×10^6 Btu/hr)	878 kwth	
Cost (10×878)		8,780
Total First Cost		$93,930
Yearly Operating Cost (Diesel-Solar)		
Capital charges ($0.15 \times 93,930$)		$14,100
Fuel cost ($10^6 \times 0.333 \times 8760 \times 0.80/10^6$) (assuming load factor = 1/3)		2,340
Total		$16,440
Yearly Operating Cost (Conventional)		
Capital charges ($0.15 \times 8,760$)		$ 1,310
Fuel cost ($3 \times 10^6 \times 0.333 \times 8760 \times 0.80/10^6$)		7,020
Total		$ 8,330

maintenance costs were comparable for both systems, and raise the breakeven price to about $1.45 per 10^6 Btu.

4.1.3 Comment on Heat Pumps

Process steam raising by pumping heat from the atmosphere is ineffective for the following reason. When the temperature difference between

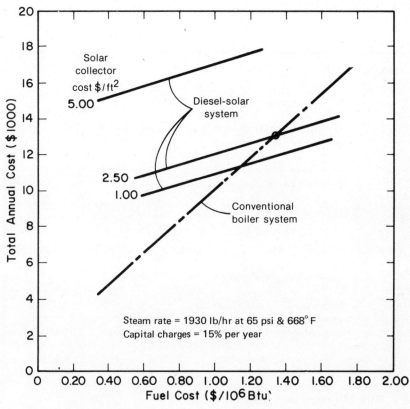

Figure 4-1-3. Diesel-solar vs. conventional steam raising costs.

materials in process and the atmosphere is greater than about 100°F, an electrically operated heat pump consumes more fuel than required in direct heating of the materials in process.

4.2 RECUPERATORS AND REGENERATORS

The effectiveness of use of fuel could be improved in many industrial processes by recovery of useful energy that is now lost as sensible heat of either exhaust gas or materials in process. Recuperators and regenerators can reduce fuel consumption by returning some of this energy to the process. They may be used to preheat the combustion air with exhaust gases.

Although from a thermodynamic standpoint recuperators and regenerators serve the same purpose, they use different heat transfer mechanisms and are applicable to different systems. A recuperator is simply a direct heat exchanger between exhaust gas and combustion air. A regenerator, on the other

Table 4-2-1.   Cost Analysis for Recuperators on Radiant Tube Heat-Treating Furnace

Additional first costs:			
- 15 Recuperators	$10,500		
- Piping, installation cost and larger blower	1,600		
- Increased engineering and design changes	250		
	$12,350		
Fuel cost $/10^6$ Btu	0.60	0.90	1.20
- Fuel saving per year	3,172	4,757	6,343
- Increased maintenance expenses for recuperator	840	840	840
Net saving per year	2,332	3,917	5,503
Rate of return on investment (10 year life and $500 salvage value)	13.9%	29.4%	43.5%

hand, is a thermal storage material, such as a slowly rotating porous disc or a stationary brick-work matrix known as a checker, which receives heat from exhaust gas and transfers it to the incoming air. The transfer is made either by means of a rotating disc in a two-channel duct, or by alternating flow of exhaust gas and combustion air through two sets of checkers.

Checkered regenerators, being refractory, are used in high-temperature corrosive environments, such as those in coke ovens, open hearth furnaces, and air-blast stoves. Disadvantages of regenerators as compared with recuperators are higher cost and loss in effectiveness through leakage from hot side to cold side. The rotating-wheel Ljungstrom air-preheat regenerator is usually applied to large volume systems at moderate temperatures (about $1200°F$) such as preheaters for boilers and gas turbines.

Recuperators are used as heat-recovery devices on certain types of processing furnaces and incinerators. Oxidation of metal-separating walls limits their application to flue-gas temperatures below about 2000°F.

Recuperators on radiant-tube controlled-atmosphere heat treating furnaces have been shown to yield a fuel saving of 23% by preheating the combustion air to 900°F using heat recovery from flue gases at 1800°F. Assuming furnace operation of 6000 hours per year, and fuel costs of 0.60, 0.90 and 1.20 dollars per million Btu, the saving in fuel cost can be compared with increased capital costs. Recuperators designed for 250,000 Btu/hr radiant tubes have recently been made available at a cost of $700 per module.[2] A summary of the cost factors for a 15-tube furnace, shown in Table 4-2-1, indicates a relatively high return on investment of 13.9% even at $0.60/$10^6$ Btu. If fuel costs double, the return on investment more than triples.

Although the above example refers to a new installation, studies have shown that recuperators may be a good investment even when fitted to existing furnaces.

4.3 BOTTOMING CYCLE ENGINE FOR WASTE HEAT RECOVERY

Many industrial processes currently reject waste heat at the relatively low temperatures of 300 to 700°F. Occasionally this heat is fed back into the process by means of recuperators, or it is used to generate steam. Often, however, no attempt is made to conserve this low grade heat, either because of marginal economics, or because process steam is not needed on the site.

An attractive method for converting waste heat to electricity or mechanical power is afforded by the organic Rankine engine used in a bottoming cycle.[3] This engine is currently under intensive development and will become commercially available in a few years. Because of their ability to use low-temperature heat sources, organic Rankine engines have great potential as stationary powerplants generating power as a by-product of industrial processes. Units may be made any size, ranging from a few horsepower to thousands of horsepower. Engine output can be coupled to electric generators, or to various other loads such as pumps, compressors, fans, conveyor drives, and mill rolls.

Gas turbines are currently used extensively in the U.S. by electric utilities to meet peak power demand and by certain industries for on-site power generation. A typical application of the bottoming-cycle engine for gas-turbine exhaust-heat recovery is shown schematically in Figure 4-3-1. Table 4-3-1 summarizes the performance of this system fitted to the GE Model PG 7791R recuperated gas turbine. Overall conversion efficiency rises from 37% to 47%, and shaft power is increased by 29% with no increase in fuel consumption.

Even higher efficiency is possible with a Diesel prime mover, producing about 500 hp per cylinder. The organic Rankine-cycle system can extract work from both the exhaust gases which are at about 700°F and from

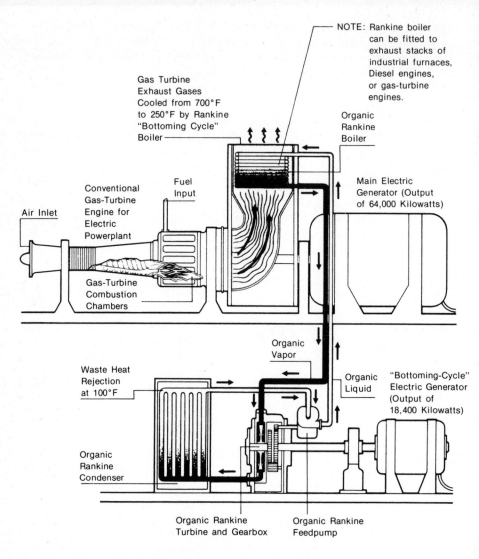

Figure 4-3-1. Schematic of organic Rankine bottoming cycle for waste heat recovery.

heat rejected in cooling cylinder walls. Efficiency of the combined Diesel and Rankine cycle is over 49%, compared with 39% for the Diesel alone.

    The organic Rankine cycle is shown in the temperature-entropy diagram of Figure 4-3-2. Superheated vapor is expanded through either a reciprocating or turbine expander (1-2). The low pressure vapor, which is superheated, passes through the gas side of a regenerator (2-3) where it transfers

Table 4-3-1.   Characteristics of Organic Rankine Bottoming Cycle
for Waste Heat Recovery From Gas Turbine

Gas Turbine [4]

Model number	GE PG7791R
Power (59°F, 14.7 psia)	64,000 kwe
SFC	0.48 lb/kwhre
Heat rate (HHV)	9,300 Btu/kwhre
η_{OA} (HHV)	37%
Exhaust gas flow rate	1.90×10^6 lb/hr
Exhaust gas temperature	715°F

Organic Rankine Cycle Bottoming Plant

Exhaust gas temperature from boiler	250°F
Fluorinol-85 flow rate	7.9×10^5 lb/hr, 1190 GPM
Turbine	
Type	Single Stage, Axial Impulse
RPM	3600
Tip diameter	6.2 ft
Blade height	6 inches
Power	18,400 kwe

Overall Plant Characteristics

Gas turbine power		64,000 kwe
Organic Rankine cycle power		18,400 kwe
Total binary plant power		82,400 kwe
% Increase in power output		29%
η_{OA}(HHV)	- Gas turbine	37%
	- Binary plant	47%
Heat rate (HHV)		
	- Gas turbine	9300 Btu/kwhre
	- Binary plant	7250 Btu/kwhre

heat to the boiler-feed liquid. The vapor leaving the regenerator is condensed (3-4-5), pumped to boiler pressure (5-6), preheated in the regenerator (6-7), and passed through the boiler (7-8-9-1).

Organic fluids have a much smaller ratio of latent heat of vaporization to sensible heat than water. Figure 4-3-3 compares the boiler temperature profiles for an organic fluid boiler and a steam boiler along with that for the Diesel exhaust gas which is the source of heat. The mean temperature difference between the exhaust gas from the Diesel engine and the vaporizing working fluid is much smaller for the organic fluid than for steam. Irreversibility is correspondingly less for the organic fluid, and the efficiency of the combined cycle is correspondingly greater.

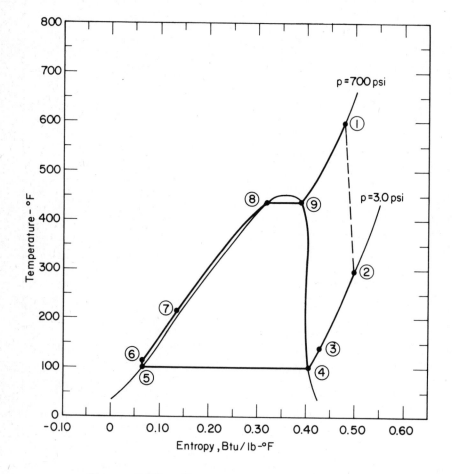

Figure 4-3-2. Temperature-entropy diagram for organic Rankine bottoming cycle engine.

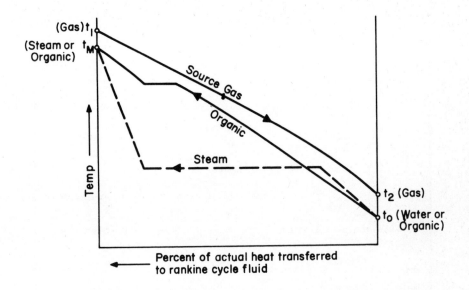

Figure 4-3-3.   Source gas and Rankine cycle temperature profiles.

The engine of an organic Rankine system is relatively simple. The organic fluid requires only a single-stage axial-flow impulse turbine, whereas steam requires 10 to 15 stages to achieve maximum efficiency. This advantage results from the higher molecular weight and correspondingly lower enthalpy drop and spouting velocity of the organic working fluid. For example, a 3600 rpm, 20,000 kwe single stage organic turbine, with the inlet at 700 psi and 600°F and the exhaust at 3 psi, will have an axial impulse wheel of 72 inch tip diameter and 7 inch blade height. This design is substantially lower in cost than the equivalent multistage steam turbine.

Because of the simple turbine design and the non-corrosive nature of organic working fluids which are compatible with carbon steel, the capital costs of organic Rankine bottoming cycles are low. Maintenance costs are also low because of the moderate temperatures and a hermetically-sealed design with integral lubrication. Table 4-3-2 shows the incremental power cost for a typical bottoming cycle plant of over 10,000 kwe that could be used for recovery of waste heat from any industrial process or engine system exhausting at 700°F. The economic advantages of the bottoming cycle are obvious: $150 per kwe for construction (which may be reasonable for multi-megawatt units) and no cost for fuel, compared to nuclear powerplants costing over $400 per kwe for construction and $0.20 per 10^6 Btu for fuel, or fossil-fuel plants costing $300 per kwe for construction and $0.50 to $1.00 per 10^6 Btu for fuel.

Table 4-3-2.   Incremental Power Cost From Rankine Cycle Portion of Binary Gas Turbine/Rankine Cycle Plant

Capital Cost Contribution		
Installed plant cost	$ 150/kwe	
Fixed charges on capital	15%	
Load factor	0. 80	
Equivalent hours/years	7000	
Capital charge contribution	0. 32	cent/kwhr
Fuel Cost Contribution	0. 00	cent/kwhr
Operation and Maintenance	0. 05	cent/kwhr
Cost of Power Leaving Plant	0. 37	cent/kwhr

REFERENCES

1. S. E. Nydick and L. J. Lazaridis, "A Jet Impingement Recuperator for Gas-Fired Radiant Tube Furnaces," presented at ASME Winter Annual Meeting, Detroit, Michigan, Nov. 11-15, 1973, ASME Paper No. 73-WA/HT-6.

2. Personal Communication with Paul Shefsiek, Holcroft and Company, Detroit, Michigan.

3. "Organic Rankine Engines for Low Exhaust Emission and Improved Energy Utilization." Thermo Electron Corporation, July 1973.

4. W. B. Wilson and W. J. Hefner, "The Role of Gas and Steam Turbines to Reduce Industrial Plant Energy Cost," General Electric Company.

Chapter Five

Fuel-Saving Methods in Selected Industries

This chapter presents specific fuel saving methods for each of six major industries—iron and steel, petroleum refining, paper, aluminum, copper and cement. Some of these methods have been discussed in Chapter 4, others are already being implemented in some industrial plants in the U.S., and others reflect practices outside the U.S.

5.1 IRON AND STEEL

The primary raw materials for the iron and steel industry are coal, iron ore, and iron and steel scrap. Iron ore and coal are first processed for use in the blast furnace where molten pig iron is manufactured. The molten pig iron is then transferred to either an Open Hearth Furnace (OHF), a Basic Oxygen Furnace (BOF), or an electric furnace where it is mixed with scrap and converted to steel. Steel is converted to slabs, blooms, and billets, primarily by the rolling of ingots. Semi-finished product from a hot rolling mill is then either formed again in a cold state, heat-treated, or forged.

5.1.1 Feedstock Preparation

The feedstock materials, iron ore and coal, must be prepared prior to being loaded into the blast furnace. Ores are converted into sinter or pellets, and coal into coke. Naturally fine ores and ore fines from screening operations, flue dust, ore concentrates, and other iron-bearing materials of extremely small particle size must be converted into agglomerates (relatively large-size pieces) because the agglomerates require a smaller amount of coke in the blast furnace.

One method of agglomeration is sintering. By means of heating to 2400–2700°F, sintering converts a mixture of fine ores, etc., into granular, relatively coarse, porous lumps that are well suited for the blast furnace. The hot sinter is cooled to about 500°F by circulating air. In most present day plants, however, the cooling air leaving the sinter machine is only at a temperature of about 250°F and therefore is of little use for heat recovery. Total fuel consumption in the sintering process is about 2×10^6 Btu/ton sinter.[1] In 1969 about 50×10^6 tons of sinter[2] (about 33% of iron ore burden) were produced.

A second practical method of agglomeration is pelletizing. It consists first of rolling ore fines into balls called green pellets by means of heating in a rotating drum or disc with moisture usually added to increase pellet strength. Then, the green pellets are fired on a travelling grate system. Fuel requirements are about 0.8×10^6 Btu/ton pellets.[1] In 1969 about 60×10^6 tons of pellets[2] (about 40% of iron ore burden) were produced. The pelletizing temperature is about 2400°F and the pellets are cooled by an updraft of air across the travelling grate. The cooling air is at temperatures comparable to those of the sintering process and, therefore, not hot enough for economical recovery of heat.

Coke is the primary fuel used in the blast furnace for the reduction of iron ore into pig iron. It is the relatively nonvolatile carbonaceous residue obtained from distillation of coal in a coke oven at 1650–2000°F. An appropriate blend of coals is charged into the oven chamber. Preheated air from regenerators enters the combustion-chamber sections of the oven where it mixes with under-firing gaseous fuel which is mostly coke oven gas. The air is preheated in regenerators by recirculated flue gas. Present U.S. practice is to quench the coke rapidly with water to prevent it from oxidizing and to cool it for transfer to the blast furnace, a process which produces dust-laden pollution and contaminated water. In European Countries (including the Soviet Union), on the other hand, the coke is cooled with a recycled inert gas from which the heat is partially recovered in steam generators and partially in inert-gas coal preheaters of incoming coal. In addition to coke, the coking process yields coke oven gas and tar products which leave the oven at 1100–1300°F. A representative fuel balance for 1969 average practice for the coking process is given in Table 5-1-1.

Estimates of available useful works of input and output materials are listed in the second column of Table 5-1-1. These data indicate that the coking process is very effective (13195/14790 = 0.9). The estimates of available useful works neglect credit for recovery of sensible heats.

5.1.2 Blast Furnace Operation

The blast furnace converts iron-bearing raw material (sinter, ore, or pellets) into pig iron for eventual conversion into steel. About 70% of the raw steel produced in the U.S. starts as pig iron and is reduced in the blast furnace, which consumes about two thirds of all the fuels used for steelmaking in the U.S. One charge of a typical blast furnace consists of 55 tons of sinter and ore, 21 tons of coke, 5 tons of slag and 1 ton of scrap, and produces about 40 tons of

Table 5-1-1.   Representative Fuel Balance for Coking Process (1969 Average Practice)

	Enthalpy Btu/lb coal	Available Useful Work Btu/lb coal
Input		
Coal	13510	13530
Underfiring gas	1280	1210
	14790	14740
Output		
Coke	10080	9580
Coke oven gas	3240	2890
Water Vapor	130	—
Tar	600	600
Light oils	130	125
Heat in stack gases	440	—
Radiation - unaccounted	170	—
	14790	13195

pig iron. The average furnace is charged about 85 times a day, or every 15—20 minutes, and yields about 3500 tons per day. The burden forms a bed which moves down the length of the furnace at an average speed of about 5 ft/min. Hot air blast at 1200°F—2000°F, injected at the tuyeres, which are large openings in the side of the furnace near the bottom, moves upward counterflow to the movement of the burden bed. Thus the charge is heated by the hot-air blast and gaseous products (blast furnace gas) of the iron ore reduction process as it moves downward and the air and gas mixture moves upward.

A representative enthalpy balance for the blast furnace is given in the first column of Table 5-1-2. Available useful works given in the second column of the table indicate that the process in the blast furnace is of relatively high effectiveness (15.1/19.1 = 0.8). In fact, the process is even more than 80% effective because the available useful work of the output has been computed without full credit for sensible heat recovery.

Table 5-1-2.   Representative Fuel Balance for Blast Furnace (1969 Average Practice)

	Enthalpy 10^6 Btu/ton of pig	Available Useful Work 10^6 Btu/ton of pig
Input		
Fuels		
Coke	15. 66	15. 70
Oil	0. 18	0. 17
Natural gas	0. 47	0. 42
Tar	0. 07	0. 07
Coke oven gas	0. 06	0. 05
Blast furnace gas for blast stoves	2. 31	2. 19
Utilities		
Steam	1. 46	0. 48
Electricity	0. 14	0. 05
		19. 1
Output		
Pig iron	8. 1	7. 7
Slag	0. 5	–
Blast furnace gas	8. 2	7. 4
Heat loss	1. 45	–
		15. 1

5-1-3 Steel Furnace Operation

Pig iron and scrap are converted into steel in furnaces of the open-hearth, the basic-oxygen, or the electric type. Although prior to 1960, the open-hearth process was predominant, by 1970 the relative production in the U.S. was as follows:[2]

Open hearth	37%
Basic oxygen	48
Electric	15
	100%

By 1978, open-hearth production is expected to be only 15% of total output.

5.1.3.1 Open Hearth Furnace

Cold limestone, iron ore, and scrap are charged into the open-hearth furnace and heated until the scrap is melted. Molten pig iron is then poured into the furnace and the mixture is heated further until the limestone reacts, forming CO_2 (the lime boil), and carbon and nonferrous metals in the pig iron oxidize. When O_2 lancing is employed to increase productivity, oxygen is injected onto the surface of the molten bath. After it reaches a temperature of 2900–2950°F, the bath is tapped to remove the liquid steel. About 43% of the fuel consumed is natural gas, 33% fuel oil, and the remainder is tar, pitch and coke oven gas. Combustion air is preheated by gaseous products to 1500–2000°F in checkered regenerators and a further 20–25% of the heating value of the fuel is recovered in boilers heated by the exhaust from the regenerators which is still very hot. Fuel requirements per ton of steel for the open hearth process are given in Table 5-1-3. It is seen from columns 1 and 3 that about 9% fuel saving is achieved through the use of O_2 lancing [(11.05–10.05)/11.05 = 0.09]. Moreover column 4 indicates that open-hearth steelmaking has an effectiveness of 6.9/8.85 or 0.78, which is a high value compared with other industrial processes.

5.1.3.2 Basic Oxygen Furnace

The Basic Oxygen Furnace (BOF) is the latest development for improved productivity of steelmaking, replacing primarily the open hearth. The BOF consumes no fuel except that required for the electrical work (61 kwhr/ton) needed for auxiliaries and for production of oxygen. Heat produced in the furnace by the oxidation of carbon, silicon and other trace materials is sufficient to bring the metal and slag to 2900°F. In fact, slag is often fed into the BOF to control its temperature.

Current practice in the U.S. is to burn the off-gases containing CO in steam-cooled or water-cooled hoods situated over the BOF vessels. The burned gases are then cooled and passed through equipment for control of air pollution. Because of the rapid cyclic operation of the BOF, it is relatively difficult to raise steam by burning waste CO. Few installations in the U.S. raise steam in this manner and those generate only a fraction of the amount normally raised in Europe and Japan where very efficient waste heat boilers have been installed over BOF's. The effectiveness of the Basic Oxygen Process is about 92%, the highest of the three steel making processes.

5.1.3.3 Electric Furnace

The percentage of steel produced in electric furnaces has gradually increased in the last ten years from about 8.5% in 1960 to about 15% in 1970. The increased use of electric furnaces is due to introduction of the Basic Oxygen

Table 5-1-3.   Representative Fuel Balance for Open Hearth Furnaces

| | Without O₂ Lancing | | With O₂ Lancing | |
	Enthalpy	Available Useful Work	Enthalpy	Available Useful Work
	10^6 Btu/ton		10^6 Btu/ton	
Input				
Fuel (natural gas, oil, tars, coke-oven gas)	3.6	3.4	1.7	1.6
Pig iron	4.5	4.2	4.5	4.2
Scrap	2.75	2.7	2.75	2.7
Utilities				
Oxygen Supply (2000 ft³/ton)	—	—	0.4	0.12
Electricity	0.2	0.07	0.7	0.23
	11.05	10.37	10.05	8.85
Output				
Raw steel	7.25	6.75	7.25	6.75
Heat	1.8	—	0.80	—
Heat loss out of stack	0.8	—	0.50	—
Steam produced	1.0	0.4	0.40	0.15
		7.15		6.9

Furnace. Because the BOF has a limited capacity for scrap, electric furnaces which can use a large proportion of scrap increased in number along with BOF's in order to process the available supply. About 500 kwhr/ton raw steel[1,2] are required in the process, or 5×10^6 Btu/ton raw steel (based on a conversion rate of 10,000 Btu/kwhr). The productivity of the furnace is increased by preheating the scrap, by operating at higher electric power levels, and by injecting oxygen. The effectiveness of electric steel making per se is about 80%. When the efficiency of generation of electricity is taken into account, however, this effectiveness becomes about 60%.

5.1.4 Slabbing, Blooming, and Billeting Operations

The traditional method of producing slabs, blooms, and billets involves a sequence of heating, rolling, and blooming operations. Liquid steel from the steelmaking furnace is poured into refractory lined ladles and then into ingot molds where solidification occurs. The ingots are brought to an equilibrium condition in soaking pits at a temperature of about 2400°F, after which they are transferred to primary rolling mills where they are rolled into slabs, blooms, or billets which are cooled and shipped to secondary mills for further shaping into plates, bars, coils, wire, tubing, etc. Prior to shaping in secondary mills, the slabs, blooms, or billets are heated to a plastic condition in reheat furnaces.

Most soaking pits are equipped with recuperators or regenerators for preheating combustion air. Fuel consumption in soaking pits varies from 0.4×10^6 to 2.0×10^6 Btu/ton of ingots.[1] Virtually all reheat furnaces have recuperators that preheat combustion air to 850–900°F. Average fuel consumption in reheat furnaces is about 2.75×10^6 Btu ton of steel.[1]

An alternative process, continuous casting, bypasses the ingot pouring and soaking pit stage of steelmaking. The saving in fuel is about 750,000 Btu/ton of steel when compared to average soaking pit practice, and about 270,000 Btu/ton of steel[4] when compared to the best soaking pit practice. The electrocity required for continuous casting is about 22 kwhr/ton[5] compared to 40 kwhr/ton[1] for ingot methods, thus resulting in an additional fuel saving of 180,000 Btu/ton of steel.

Another significant advantage of continuous casting is that the yield of semifinished products is about 95% compared to 80% for the ingots. This represents considerable fuel saving in steelmaking, about 60 kwhr or 0.6×10^6 Btu/ton of semifinished product, because 12% of the steel produced does not have to be reprocessed as scrap in the electric furnace.

In summary, continuous casting could result in fuel saving between 1×10^6 and 1.5×10^6 Btu/ton of steel.

5.1.5 Heat-Treating and Forging Operations

Heat-treating furnaces serve a number of different purposes such as annealing, normalizing, preparation for forging, tempering, etc. The fuel required for these furnaces is supplied as electricity, oil, natural gas, or by-product gases. In annealing operations, steel is heated to 1200–1500°F to soften the material and obtain a microstructure with lower internal stress. Average fuel consumption is about 4.5 x 10^6 Btu/ton.[6] About 25% of U.S. steel production is heat treated, and about 15% is processed in forging furnaces.

5.1.6 Overall Effectiveness

Iron ore consists primarily of two oxides, magnetite (Fe_3O_4) and hematite (Fe_2O_3). For the conversion of ore to steel the minimum amount of fuel required is given by the change in available useful work. If the conversion occurs at room temperature, the change in available useful work is also equal to the change in Gibbs free energy at 298°K. For magnetite the change in Gibbs free energy is 5.69 x 10^6 Btu/ton[1] and for hematite 6.28 x 10^6 Btu/ton.[1] When trace elements in steel such as C, Mn, P, S, etc. and the relative proportions of magnetite and hematite are included in the calculation, then the weighted change in Gibbs free energy is 6 x 10^6 Btu/ton raw steel.

In practice, the change in available useful work is supplied by a number of fuels used either directly in the reduction process or indirectly in the various pieces of machinery. To evaluate the overall effectiveness of steel making, a flow chart of available useful work based on average data for 1969 is given in Figure 5-1-1. Heavy solid lines indicate flow of primary fuels such as coal, oil, and natural gas, lighter solid lines indicate steam or electricity, dotted lines show the flow of by-product fuels such as coke oven gas or blast furnace gas, and wavy lines indicate the flow of iron and steel including scrap. The chart indicates that the total input of available useful work is 28.6 x 10^6 Btu/ton of steel consisting of 25.8 x 10^6 Btu/ton in the form of primary fuels and 2.8 x 10^6 Btu/ton in the form of scrap, and the output is 6 x 10^6 Btu/ton of steel. Thus the overall effectiveness is 6/28.6 = 21%. Of the approximately 23 units of loss of available useful work, the chart indicates that about 5 units are lost in electricity and steam generation, 5 units in pig iron making, 2 units in steel making, 7 units in steel processing, and 3 units in unused by-product fuels. These results suggest significant opportunities for fuel saving.

5.1.7 Fuel Needs

Figure 5-1-2 shows the average distribution of fuel consumption among the various operations involved in steel making. It is seen from the figure that about 73% of the total fuel (19.2 x 10^6 Btu/ton steel) is supplied by non-petroleum sources, primarily coal, and 27% by petroleum-based fuel. The 27% figure is well below the 60% average for industry in general, and thus, the steel industry uses an unusually large proportion of domestic fuels.

5.1.8 Recommendations for Fuel Savings

Because steelmaking uses so much fuel, a fractional improvement justifies substantial capital investment. Such investment, particularly in basic

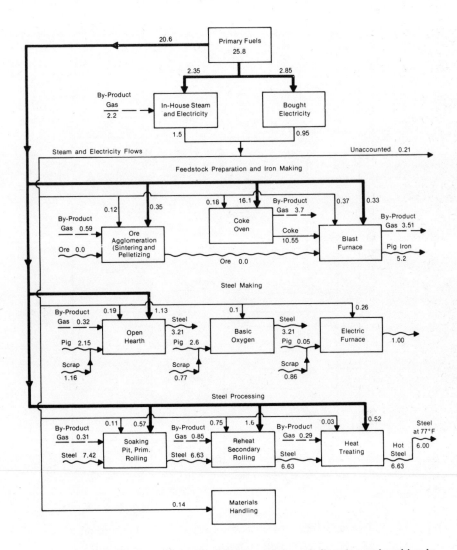

Figure 5-1-1. Average available useful work flow in steel making in 1969. All quantities in 10^6 Btu/ton of steel. The dashed lines denote by-product coke-oven and blast-furnace gases.

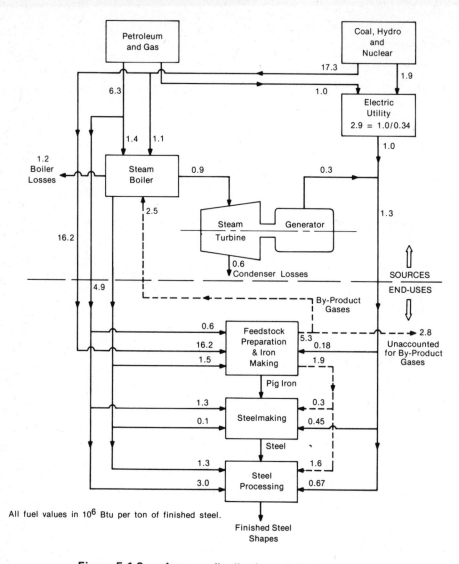

Figure 5-1-2. Average distribution of fuel consumption in U.S. steel industry in 1968-1969.

oxygen furnaces, has resulted in a gradual decline in specific fuel consumption over the past decade. From 1960 to 1968, the specific fuel for steel production declined at an average rate of about 1.7% per year. A detailed review of present steelmaking practices in the light of comparisons with the practices of other countries and the discussion of thermodynamic availability in Sec. 5.1.6 indicates that further substantial fuel savings are possible within the limits of existing technology. The specific fuel consumption can be reduced from 26.5×10^6 Btu/ton to about 17.2×10^6 Btu/ton of steel if the following steps are taken:

1. Increase the ratio of iron-ore pellets to sinter in the charge entering the blast furnace. The proportion of sinter can be reduced from 35% to 20%.
2. Reduce sensible heat losses from sinter, pellets, coke, and coke oven gas by charging hot materials directly into blast furnace.
3. Preheat combustion air supplied to sinter and pellet furnaces.
4. Increase the air blast temperature (from 1600 to 2000°F) and the top gas pressure (from 2 to 15 psig) in the blast furnace: in Japan coke consumption is 20% less than in the U.S. because of higher air blast temperatures and higher top-gas pressures.[7,8]
5. Use more coke-oven gas to replace natural gas in blast furnace. Currently, some by-product gases are wasted.
6. Continue the replacement of open hearth furnaces by basic oxygen furnaces so that BOF production is increased from 43 to 60% of industry output. Also, use off-gases from BOF more effectively.
7. Increase use of oxygen injection in open-hearth and electric-furnace steel making.
8. Improve regenerators of open-hearth furnaces.
9. Use continuous casting methods more and ingot methods less. Continuous casting can be increased from 5 to 50% of steel making.
10. Improve recuperators for ingot soaking pits, reduce soaking times by tightening control of process flow, and eliminate reheating for secondary rolling.
11. Combine generation of process steam and electric power wherever possible.
12. Use by-product gases rather than coal or petroleum for generation of steam.

Figure 5-1-3 shows the flow of available useful work for steelmaking if the above steps are implemented. The overall effectiveness is 30.5% (6/19.7), compared to the average effectiveness for 1969 of 21.0%. The breakdown of the projected reduction in loss of available useful work into the various steps of steelmaking can be seen by comparing Figure 5-1-1 with Figure 5-1-3.

Figure 5-1-4 shows the distribution of fuel consumption for steelmaking that results if the above steps are implemented. It is estimated that the fraction of fuel supplied by petroleum sources will be reduced from 27 to 18% of industry needs.

Even further long-term fuel savings are possible. Within about 20 years, it is reasonable to assume that the basic oxygen furnace (or even more advanced methods) will provide at least 75% of capacity, the remainder being electric-furnace capacity for scrap steel. Coke consumption should decline to about 0.375 lb per lb of pig iron, and continuous casting methods may be

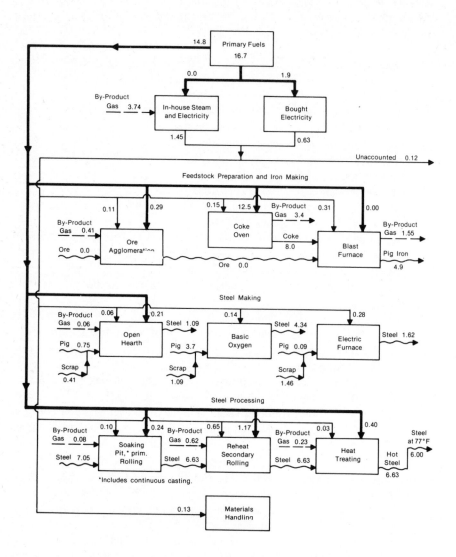

Figure 5-1-3. Potential available useful work flow in steel making. All quantities in 10^6 Btu/ton of steel. The dashed lines denote by-product coke oven and blast furnace gases.

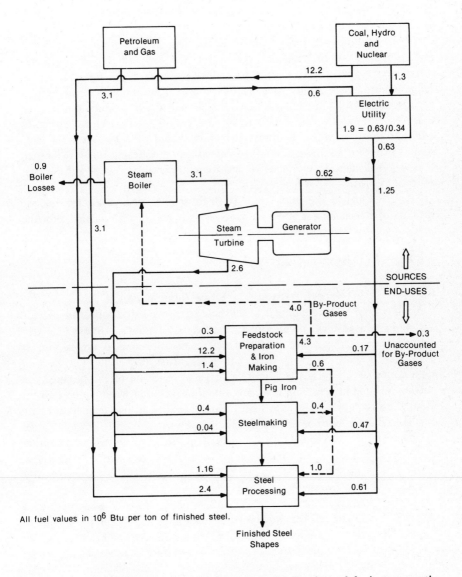

All fuel values in 10^6 Btu per ton of finished steel.

Figure 5-1-4. Potential average distribution of fuel consumption for improved steel industry practices.

applied almost universally. These changes, along with general adoption of highly effective recuperators, could reduce average fuel requirements to about 14 x 10^6 Btu/ton of raw steel, or a little more than half the 1968 figure.

Considerable development work has been performed on direct reduction methods of steelmaking. However, the complete or partial replacement of the blast furnace with pre-reduced ore from the proposed direct reduction processes results in a net fuel consumption more than the current average. This results primarily from a decrease in by-product production of coke oven gas, tars, coke breeze, and blast furnace gas for use in steel processing. In addition, the conversion to direct reduction would generally result in large requirements for natural gas, a rather scarce fuel domestically. Direct reduction, however, will increase productivity and therefore could result in a lower cost product, but with greater fuel requirements.

Finally, the application of high temperature gas-cooled nuclear reactors to steelmaking is under study. Success in this direction could alter the preceding considerations for the longer term.

5.2 PETROLEUM REFINING

5.2.1 Process Description and Fuel Consumption

Petroleum refining is a very heterogeneous industry because of wide variations in feedstock, product mix, and process methods. Although no two refineries are exactly alike, certain process similarities exist. A typical process flow chart is shown in Figure 5-2-1. Crude oil is first distilled in a unit fired by natural gas or oil at nearly atmospheric pressure. Some of the products are further distilled in a vacuum to separate components of higher molecular weight from those of lower. Heavy distillates which vaporize at high temperatures are removed at the bottom of the distillation tower, and lighter distillates are removed at the top. The light distillates are then further separated by adsorption and absorption. The resulting straight-run products are crude natural gas, gasoline and naphtha. Most other straight-run products are subsequently desulfurized or hydrogenated.

The fractions having high boiling points are used as feedstock to thermal or catalytic crackers where hydrocarbon chains are decomposed so as to increase the yield of gasoline and light oils such as kerosene and heating oils. The residue that floats to the bottom of the distillation tower is processed by viscous breaking, catalytic cracking, hydrocracking, lube-oil manufacture, or asphalt manufacture. It may also be transformed into gasolines and light oils by sufficient processing. For example, high octane gasolines are produced by processing the catalyzed residue in polymerization and alkylation units.

In general, the fuel consumption of a refinery is dependent on its complexity and age. As the complexity of a refinery increases to produce a greater variety of products, fuel consumption will increase. Also, older refineries

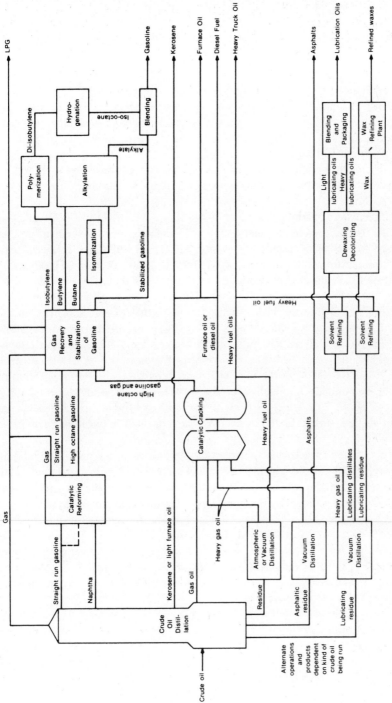

Figure 5-2-1. Typical petroleum industry process flow.

that were designed when fuel was less expensive usually consume more fuel. Over the past 20 years, the fraction of gasoline output has risen slowly from about 42% to 50% of total refinery production but average fuel consumption in U.S. refineries has remained nearly constant in the range from 4.4×10^6 to 4.8×10^6 Btu per ton of refined product. One the average, about 11.5% of the total heating value of crude petroleum feedstock is consumed in the refining operation. In 1968, petroleum refineries in the U.S. consumed 2.6×10^{15} Btu of fuel for a total 590×10^6 tons of product.

The Nelson Complexity Factor, originally conceived by W. L. Nelson as a guide for the estimation of process capital costs, has also been suggested as a tool to correlate fuel consumption with product mix and yields, and to compare the efficiencies of various typical refineries. Employing data in the Oil and Gas Journal 1973 Annual Refinery Survey[9] and Nelson's individual process complexity factors,[10] calculations were made for 28 Gulf Coast refineries with fuel consumption values from 2.3×10^6 to 6.15×10^6 Btu/ton of products.[11] In Figure 5-2-2, the fuel consumption of product for these refineries is plotted versus the refinery complexity factor. The fuel consumption of about eight refineries falls below the high efficiency line,[12] that of ten falls between the high and low efficiency lines,[12] and that of the remaining ten exceeds the low efficiency line. Analysis of the data presented in Figure 5-2-2 shows that about 24% of the fuel consumed in a refinery can be conserved if all of the refineries were brought to at least the high efficiency level by the better application of existing technology. Elimination of the three high data points still results in a saving of about 15%, which agrees with the estimate of Whitcomb and Orr[12] for potential fuel saving in the next five years through application of existing technology.

Table 5-2-1 shows the distribution of fuel for the petroleum-refining industry in 1967. Although most of the fuel is consumed in direct heating and in steam generation for which coal is suitable, almost all of the industry's needs (over 90%) were satisfied by consumption of petroleum and natural gas.

It is difficult to establish a detailed accounting of fuel flow in refinery operations because oil Companies regard their process data as proprietary. Nevertheless, it can be stated that U.S. refinery practices have been optimized in favor of minimum capital investment rather than fuel conservation because of extremely low fuel prices (as low as 20 cents per million Btu for natural gas in the Gulf Coast area).

Although detailed data are lacking, we have estimated the effectiveness of fuel utilization in refining by considering the information reported about the Pascagoula refinery of the Standard Oil Company.[13] The enthalpy balance for this refinery is listed in the first column of Table 5-2-2. Estimates of the available useful works of the input and output materials are listed in the second column. Two conclusions can be deduced from these estimates. From the overall point of view, the effectiveness of refining is extremely high, namely

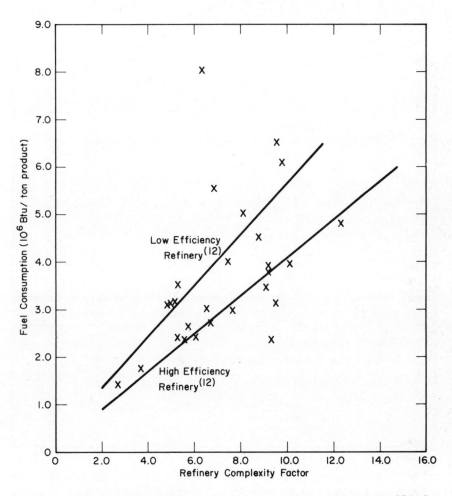

Figure 5-2-2. Fuel consumption vs. complexity factor for 28 U.S. petroleum refineries.

5.76/6.36 = 0.91. On the other hand, from the point of view of utilization of natural gas and other fuels, the process is extremely ineffective because for a mere increase of 5.76−5.7 = 0.06 in available useful work from crude to petroleum products, an amount of 0.23+0.43 = 0.66 in available useful work of fuels is consumed, yielding an effectiveness of fuel utilization equal to 0.06/0.66 = 0.09.

5.2.2 Recommendations for Fuel Savings

The large differences in specific fuel consumption between refineries as well as consideration of available useful work just cited indicate that present

Table 5-2-1.   Distribution of Fuel for Petroleum Refining in 1967-1968

Sources of fuel	10^6 Btu per ton of output product
Petroleum and by-products (residual oil, refining gas, etc.)	2.06
Natural gas	1.93
	3.99
Coal and other fuels	0.16
Electricity (at 10,000 Btu/kwhr fuel equivalent)	0.30*
	4.45

*Approximately 35% self-generated.

End-use	Percent
Direct firing	60
Process steam for heating and hydrogenation	26
Steam for mechanical drives	8
Electricity	6
	100

refining practices consume far more fuel than required for the task. This conclusion suggests that an extensive and detailed study of refinery processes would show exactly how major fuel savings can be made.

Even without further study, substantial reduction in fuel consumption can be attained by means of the following procedures:

- Increase generation of by-product electricity in conjunction with generation of process steam regardless of the electrical needs of the refinery.
- Increase recuperation for all combustion systems used in direct heating.
- Improve efficiency of steam boilers and reduce steam losses in stand-by equipment.
- Increase heat interchange between process streams, and feedstock streams of different units.
- Use hydraulic turbines to recover mechanical power from the return to low pressure of high-pressure product streams.

Table 5-2-2.   Fuel Balance for the Pascagoula Refinery of the Standard Oil Company

	Enthalpy 10^6 Btu/barrel	Available useful work 10^6 Btu/barrel
Input		
Crude	6. 00	5. 7
Natural gas	0. 24	0. 23
Fuel	0. 45	0. 43
	6. 69	6. 36
Output		
Petroleum products	6. 06	5. 76

- Substitute more plentiful fuels, such as coal, for natural gas in direct-fired process heaters.

   The average process steam pressure, exclusive of steam used for mechanical drives, is less than 200 psi. About 26% of fuel consumption in the refinery, 1.14×10^6 Btu/ton of product, is in the form of process steam. As discussed in Section 4, this steam can generate about 80 kwhr of electricity per ton of product at the expense of an incremental fuel consumption of 0.27×10^6 Btu/ton. Electrical requirements of the refinery are about 28 kwhr per ton so that 52 kwhr per ton can be sold to utilities. At 1968 production levels, this amounts to 31 million kwhr, or an average continuous power level of 3550 megawatts. Since purchased electricity amounts to about 18 kwhr per ton, the net saving would be 70 kwhr per ton (0.7×10^6 Btu/ton) minus the additional fuel consumption (0.27×10^6 Btu/ton). Overall, this is a saving of 0.43×10^6 Btu per ton (0.25×10^{15} Btu per year in 1968).
   Additional fuel savings are obtainable through the use of air preheat on combustion burners. If only half the direct-fire heat applications in the refinery are fitted with recuperators of 15% efficiency, the reduction will be

$$\frac{1}{2} [0.60 \times 4.4 \times 10^6] (0.15) = 0.20 \times 10^6 \text{ Btu per ton.}$$

This improvement, together with the savings realized by generating more by-product electricity, will reduce total fuel needs by 14%. Considering the additional changes previously noted, a reasonable goal for the early 1980's would be a saving of 25% and an average specific fuel consumption of 3.3×10^6 Btu per ton of output. Ultimately, of course, basic process changes could yield even greater improvement.

Consumption of petroleum products rose by an average of 3.7% per year throughout the 1960's. If this growth in demand is not curtailed by rationing or other methods, the 1980 consumption will be about 920×10^6 tons. If the recommended improvements could be implemented by then, fuel consumption for refining will be 3×10^{15} Btu, or 17% above the 1968 requirement.

5.3 PAPER AND PAPERBOARD

5.3.1 Process Description and Fuel Consumption

The primary raw material in all paper products is cellulose derived from wood. Quantities of waste paper and other fibrous materials such as rag are also used, but to a lesser degree. Although the total amount of recycling has increased, recycled waste paper as a percentage of annual paper production has shown a gradual decline[14,15] over the past 15 years, dropping from 28% in 1958 to 21% in 1967 and 19% in 1972.

Paper manufacture consists of two major operations, namely making pulp from wood, and making paper from pulp. Both operations consume fuel for heating and for mechanical drives, the largest fuel fraction being consumed by pulping processes.

In the pulping operation, bark, which is low in fiber content, is first removed from logs by mechanical means. The logs are then converted into pulp, either by mechanical grinding or by means of a chemical cooking process which removes the ligneous constituents of the wood. Chemical pulping methods, primarily the sulfite and sulfate processes, represented about 85% of the new pulp produced in 1967 and 90% in 1972. For 1972 the raw materials for the industry were:

Recycled waste paper	19%
Ground wood pulp	8%
Chemical pulp	73%
	100%

The paper-making operation, wherein finished products are produced from the pulp, consists of four major steps: stock preparation, sheet formation, water removal, and sheet finishing. Fuel is needed for pumps and mechanical drives and for drying operations.

Production of paper and paperboard[14,16] totaled 47 x 10^6 tons in 1967, and 59 x 10^6 tons in 1972. If this growth rate were maintained, production would amount to about 85 x 10^6 tons by 1980.

Depending upon how one accounts for the fuel value of certain waste products that result from pulping operations, two values can be calculated for the average fuel consumption. In 1967 the paper industry consumed 1.0 x 10^15 Btu of purchased fuel and 2.3 x 10^10 kwhr of electricity or a total of 1.13 x 10^15 Btu based on 10^4 Btu/kwhr.[17,18] Since in 1967 the fuel consumption in the industrial sector was 22.8 x 10^15 we would conclude that papermaking represented a percentage of industrial fuel consumption equal to

$$100 \; \frac{1.23 \times 10^{15}}{22.8 \times 10^{15}} = 5.4\%.$$

This result is misleading, however, because pulping processes yield substantial quantities of fuel-equivalent materials in the forms of wood bark and spent pulp liquors which are used in boilers.[19,20] In 1967 by-product fuels of this type produced an additional 0.68 x 10^15 Btu of heat that was used in paper making. Moreover, had the waste product reclamation throughout the industry reached the level of the most efficient plants then in operation, the heat recovery would have been more than 25% greater, or about 0.85 x 10^15 Btu.[17] Since burning of the waste products represents consumption of fuel resources that could be used for a variety of purposes other than paper manufacturing, it is reasonable to charge the paper industry with this added fuel consumption. Thus, a more accurate value for the fuel consumed for paper production in 1967 would be 2.08 x 10^15 Btu or about 9% of the industrial fuel consumption. The figure of 9.0 may be an overestimate, because other industries may also derive fuel from by-products generated in the corresponding manufacturing processes. Nevertheless, it is clear that pulp and paper manufacturing represent a fuel-intensive activity, with each ton of paper requiring the equivalent of about one ton of fuel oil.

A preliminary study of paper manufacturing processes indicates that important fuel savings can be accomplished by means of existing technology. Specific recommendations are given below. The average specific fuel flow in the paper industry is approximately as shown in[14,15,17,18,21] Figure 5-3-1. The data in the figure have been derived from the data reported for total paper production and fuel consumption in 1967. They can be summarized in terms of fuel resources and end use as shown in Tables 5-3-1 and 5-3-2.

The data reveal that in 1967 paper-making processes were relatively effective in conserving scarce fuels such as gas and petroleum. Approximately 50% of the purchased fuels were in the form of coal. Moreover, if it is assumed that electricity purchased from utilities is generated by using the average mix of 64% coal, hydro, and nuclear and 36% gas and petroleum, then the proportion

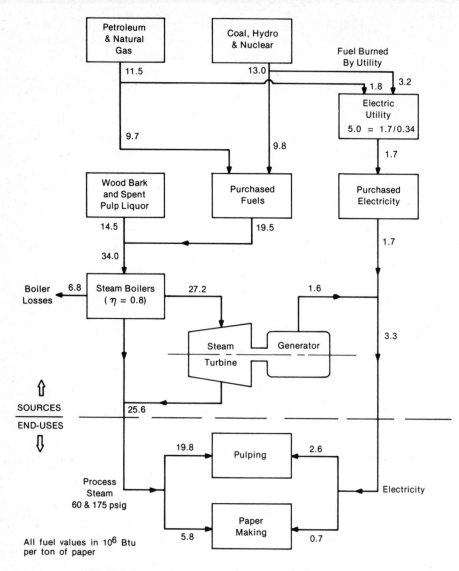

Figure 5-3-1.   Average distribution of fuel for U.S. paper industry in 1967.

of gas and petroleum-based fuels used in paper-manufacturing was only 30%, a fraction substantially smaller than the 60% fraction for industry in general. Unfortunately a marked shift away from coal and towards gas and petroleum-based fuels occurred recently. By 1973, coal accounted for 21% and gas and petroleum-based fuels for 79% of purchased fuels.

Table 5-3-1.   Fuel Sources for Paper Industry[17,22]

Waste Products	Purchased Fuels	Purchased Electricity*
37%	50%	13%

*Assuming an average of 10,000 Btu per kwhr for electricity generation.

Table 5-3-2.   End-Use of Fuel by Form and Process for Paper Industry[14,17,19]

	Pulping	ᵻPaper Making
Steam (includes boiler loss)	58%	17%
Electricity**	20%	5%

**Includes that portion generated internally by paper mills.

5.3.2 Recommendations for Fuel Savings

The fuel consumption of a given paper mill will differ from the averages shown above because of, for example, the type of paper produced, the type of wood used as feedstock, and the degree to which new techniques and processes are incorporated in the plant design. Nevertheless, the average fuel-flow data suggest procedures for saving fuel. Some of these procedures are as follows:

1. Incorporate paper-forming processes which require less fuel for drying; specifically, Thermo Electron's Lodding K-Former[23] can reduce the water throughput by as much as a factor of 4, and the fuel demand of the paper-making phase of production by 55%. Most of the saving is due to the reduced drying requirements that result because of less stratification in the wet sheet—more water can be pressed from the sheet prior to entering the dryer section.
2. Increase the integration of pulping and paper-making operations to eliminate fuel used for pulp drying prior to shipment from the pulp mill to the paper mill, which requires 3.4×10^6 Btu per ton of pulp dried.
3. Increase use of continuous rather than batch digesters. Continuous digesters reduce fuel consumption by 1.3×10^6 Btu per ton.

4. Increase recovery in waste heat boilers of fuel equivalent in bark and spent-pulp liquor; the 1967 industry average can be improved by at least 25%, from 14.5×10^6 to 18.0×10^6 Btu per ton of paper.[17]

5. Exploit all opportunities to produce electricity prior to generating low-temperature process steam, and sell surplus electricity to utilities.

6. Increase recycling of waste paper. The industry trend throughout the 1960's was toward less recycling, even though recycled paper requires less than 1/4 as much fuel as paper made from raw wood product.*

Even neglecting the effects of increased recycling, it is possible to achieve substantial fuel savings by employing the other procedures noted above. Figure 5.3.2 shows the approximate fuel flow for a pulp and paper production process which uses these procedures fully. The results are also summarized in Table 5.3.3. It is seen from the table that a major reduction in externally supplied fuels is accomplished. Fuel from waste products rises from 14.5×10^6 to 18.0×10^6 Btu per ton of paper, but purchased fuel drops from 19.5×10^6 to 11.7×10^6 Btu per ton. Electricity purchases are not only eliminated entirely, but the paper industry becomes a net producer of electricity. The electricity that is sold can be accounted for in terms of equivalent fuel savings by the utility industry amounting to $2.0 \times 10^6/0.34$ or 5.9×10^6 Btu per ton of paper.

Thus, total fuel consumption drops to 61% of the 1967 level (23.8×10^6 versus 39.0×10^6 Btu per ton); but, more importantly, the fuel required from sources external to the paper industry is reduced to less than 25% of the previous level. Excess electrical output of the paper industry would be 585 kwhr per ton of paper produced. At 1968 production levels (50×10^6 tons per year) this amounts to 2.9×10^6 kwhr, or a continuous power level of 3330 megawatts.

If the improved specific fuel consumption values listed in the second column of Table 5-3-3 are applied to a typical paper mill producing 1000 tons per day, the plant generating process steam and electricity would have the following characteristics:

- Process steam flow   630,000 lb/hr
                            (0.85×10^9 Btu/hr)
- Electricity
  − Consumed by mill   35.0
  − Sold to utility   24.4
                            ─────
                            59.4 megawatts

*If waste paper were used as a fuel, then waste paper recycling would not be a desirable fuel saving method because the fuel equivalent of paper is about the same as the fuel saving in making paper through recycling.

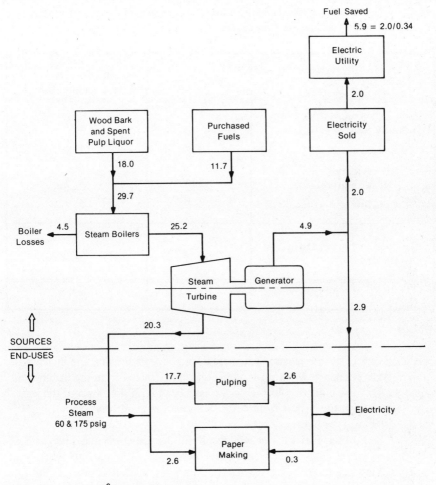

All fuel values in 10^6 Btu per ton of paper.

Figure 5-3-2. Fuel distribution for U.S. paper industry. Improved processes.

Since all fuel is intially used in steam raising, there would be considerable flexibility in the choice of fuel for such a dual plant. Therefore, coal should be given priority over gas or oil, even though greater expenditures for pollution control equipment would probably be incurred.

An interesting aspect of the dual plant is its ability to operate efficiently over widely varying production rates for the paper mill that it serves. For example, if the mill were completely shut down, the electric powerplant could still be fully employed and the turbo-generator could again run at its

Table 5-3-3.   Specific Fuel Consumption in Paper Industry for 1967 and Under Conditions of Improved Fuel Management

	Average Specific Fuel *	
	1967/1968	Improved Fuel Management
Net fuel consumption from outside sources		
Heat	19. 5	11. 7
Electricity (fuel equivalent)	5. 0	-5. 9
Total outside fuel	24. 5	5. 8
Fuel consumption from waste products	14. 5	18. 0
TOTAL	39. 0	23. 8

*Million Btu per ton.

design rating of 59 megawatts. This coincidence occurs because the increase in power cycle efficiency when no process steam is extracted from the turbine will almost offset the reduction in fuel input associated with shutdown of the waste recovery boilers. Fuel input from external sources rises slightly, from 4.9×10^8 to 5.9×10^8 Btu per hour.

Further fuel savings will result in the paper-making operation if dry-process techniques are successfully developed. Experiments with dry paper-making are now being carried out in Scandinavian countries, but no method is at present commercially successful. Dry forming processes could ultimately eliminate the drying requirements entirely and cut fuel consumption by about 2.6×10^6 Btu per ton below even that of the K-Former system. Other possible long-range developments include substitution of alternative fiber sources in place of wood pulp.

5.3.3 Estimate of Available Useful Work for Pulping

The ultimate potential for fuel saving in the paper industry can be evaluated by calculating the available useful work required for the separation of cellulose and hemicellulose from wood. This work cannot be calculated exactly because of the complexity of the chemical reactions involved in the separation process and because of the lack of thermodynamic data. Nevertheless, we have established an approximate value by considering the sulfate alkaline process and making some reasonable simplifications. The sulfate process was considered because 90% of pulp is obtained by an alkaline process.

On a dry basis, the chemical composition of wood has been assumed to be 60% cellulose, 12% hemicellulose, and 28% lignin.[24] In the sulfate process, a mixture of sodium hydroxide and sodium sulfide is used for the delignation. These chemicals act by degrading the insoluble cross-linked lignin through the breaking of bonds within the lignin macromolecule; then the fragmented lignin is solubilized and removed from the individual fibers. The overall reaction can be represented by the material balance[24] shown in Figure 5-3-3. This reaction is followed by the recovery operations shown in Figure 5-3-4. These operations involve the following steps: (a) burning of black liquor solids (with sodium

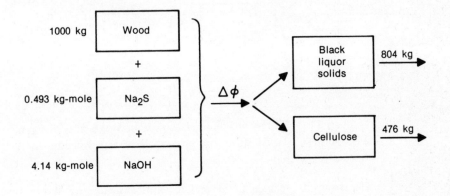

Figure 5-3-3.   Sulfate alkaline pulping reactions.

(a) $\boxed{\text{Black liquor solids}}$ $+ O_2 = CO_2 + SO_2 + H_2O + Na_2CO_3$

(b) Dissolution of smelt

(c) $Na_2CO_3 + Ca(OH)_2 = CaCO_3 + 2NaOH$

(d) Filtration of $CaCO_3$

(e) $CaCO_3 = CaO + CO_2$

(f) $Na_2SO_4 + 2C = Na_2S + 2CO_2$

Figure 5-3-4.   Recovery operations for sulfate alkaline pulping.

sulfate added, step f); (b) dissolution of the smelt to recover the sodium carbonate (Na_2CO_3); (c) caustizing of the Na_2CO_3 solution to recover sodium hydroxide (NaOH) and obtain insoluble calcium carbonate ($CaCO_3$); (d) filtration of $CaCO_3$ from the solution; (e) calcination $CaCO_3$ in a kiln; and (f) reduction of sodium sulfate (Na_2SO_4) into sodium sulfide (Na_2S) (step a), and replenishment of sulfur lost as sulfur dioxide (SO_2) and sulfur salts.

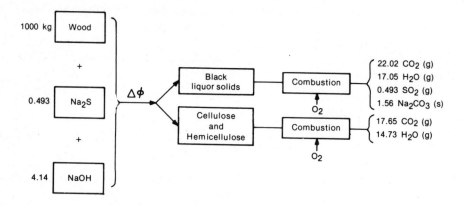

Figure 5-3-5.   Sulfate alkaline pulping reaction plus combustion of reaction products. All quantities in kg-moles except when otherwise stated.

The change in available useful work $\Delta\Phi$ during the reaction shown in Figure 5-3-3 is difficult to calculate directly because of lack of thermodynamic data for complex organic molecules. To avoid the difficulty, the output materials in Figure 5-3-3 were subjected to combustion assuming that the C:O:H: ratio for hemicellulose is the same as that for cellulose. The combustion products are shown in Figure 5-3-5. They can also be produced by direct combustion of wood plus the reactions indicated in Figure 5-3-6. For all input and output materials at room temperature, combining the reactions in Figure 5-3-5 and 5-3-6 into a cycle permits the calculation of the changes in enthalpy ΔH and available useful work $\Delta\Phi$ (here also equal to the change in Gibbs free energy ΔG) for the sulfate alkaline pulping process. The results are

$$\Delta H = -0.2 \times 10^6 \text{ Btu/ton dry wood}$$

and

$$-0.2 \times 10^6 < \Delta\Phi < 0.1 \times 10^6 \text{ Btu/ton of dry wood.}$$

The two limiting values for $\Delta\Phi$ are the result of approximations that had to be made and that can be summarized as follows:

Partial reactions

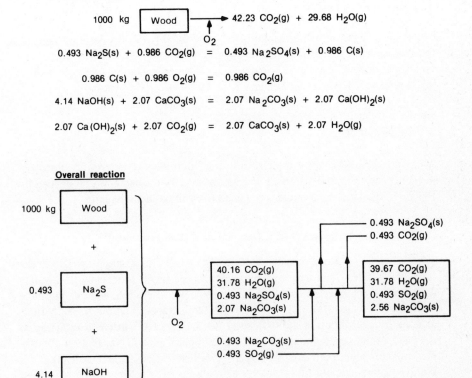

$$0.493\ Na_2S(s) + 0.986\ CO_2(g) = 0.493\ Na_2SO_4(s) + 0.986\ C(s)$$

$$0.986\ C(s) + 0.986\ O_2(g) = 0.986\ CO_2(g)$$

$$4.14\ NaOH(s) + 2.07\ CaCO_3(s) = 2.07\ Na_2CO_3(s) + 2.07\ Ca(OH)_2(s)$$

$$2.07\ Ca(OH)_2(s) + 2.07\ CO_2(g) = 2.07\ CaCO_3(s) + 2.07\ H_2O(g)$$

Overall reaction

Figure 5-3-6. Alternate method of creating the combustion products in Figure 5-3-5. All quantities in kg-moles except when otherwise stated.

1. Enthalpies and available useful works of formation of inorganic materials were obtained from standard thermodynamic tables[25] except for the available useful work of Na_2S which was assumed equal to 0.94 of the tabulated standard enthalpy.
2. The enthalpy ΔH_c of cellulose was estimated from heats of combustion;[26] the available useful work $\Delta\Phi_c$ of cellulose was estimated by means of van Krevelen's method of group contributions[27] and polymerization thermodynamic data.[28]
3. From the cycle resulting from the combination of the reactions in Figures 5-3-5 and 5-3-6 the change in enthalpy ΔH for the sulfate alkaline pulping process was found to be -0.2×10^6 Btu/ton of dry wood.

4. The ratios $\Delta\Phi_w/\Delta H_w$ for wood and $\Delta\Phi_1/\Delta H_1$ for lignin were considered as adjustable parameters.

Thus, from the reactions in Figure 5-3-3 (or the cycle resulting from the combination of the reactions in Figures 5-3-5 and 5-3-6) it was found that the change in available useful work $\Delta\Phi$ for the sulfate alkaline pulping process is in the range between -0.2×10^6 and 0.1×10^6 Btu/ton dry wood.

5.4 ALUMINUM

Aluminum is produced by the electrolysis of aluminum oxide, Al_2O_3, dissolved in molten cryolite, $AlF_3 3NaF$, a process essentially the same as that devised by Hall in 1886. Aluminum oxide is obtained from the mineral bauxite.

5.4.1 Process Description and Fuel Consumption[29]

Bauxite ore contains 40% to 60% Al_2O_3 (alumina) by weight in the form of hydrates. It is separated by means of the Bayer process which involves digestion of bauxite in a hot caustic soda solution, removal of insoluble impurities, precipitation of $Al_2O_3 3H_2O$, and transformation to anhydrous alumina by calcination at 2200°F. Fuel for the process, consumed primarily in the calcination step, amounts to 10×10^6 Btu per ton of Al_2O_3 or 19×10^6 Btu per ton of aluminum metal. Cryolite, which melts at about 1800°F, at slightly higher temperatures dissolves alumina up to as much as 10% to 20% of its weight, with a resultant decrease in melting point. Electrolysis is carried out in carbon-lined boxes into which carbon rods project. An electrical potential is applied so that the box serves as the cathode and the rods as the anode. Upon electrolysis the alumina is decomposed; the aluminum metal is deposited at the cathode in a molten condition (about 1800°F) and the oxygen is deposited at the anode. The liquid aluminum, being heavier than cryolite, sinks to the bottom of the vessel and is removed by tapping. The carbon anode is oxidized by the electrolysis to CO_2.

Part of the aluminum reduced by electrolysis is reoxidized by the CO_2, an inverse process which reduces the plant efficiency and produces an effluent mixture of gaseous CO and CO_2.

Considerable variations in the electric power requirements for primary aluminum production exist from plant to plant; typical numbers for production cells range from 13,600 to 16,400 kwhr per ton of aluminum. The process also consumes about 1000 pounds of carbon and about 100 pounds of cryolite per ton of aluminum. Other fuel-consuming processes in the aluminum industry are foundry melting and heat-treating operations such as homogenizing, annealing, solution treating, and artificial aging. These processes consume an average of about 7×10^6 Btu per ton of aluminum, almost all supplied by natural gas or petroleum. Thus, the average amount of fuel per ton of primary aluminum is 190×10^6 Btu/ton.

Primary aluminum production in the U.S. increased from 2.0×10^6 tons in 1960 to 3.25×10^6 tons in 1968. Assuming an average electrical demand of 15,000 kwhr per ton, the electrical consumption for the aluminum industry amounted to 4.9×10^{10} kwhr in 1968, or about 3.7% of total U.S. electrical output. Total fuel requirement for aluminum production, primary and scrap, was 0.63×10^{15} Btu, or about 2.8% of U.S. industrial fuel consumption.

With both feed materials and products being at room temperature, the minimum amount of available useful work and, therefore, the minimum amount of fuel required for the reduction of Al_2O_3 into aluminum is 376,800 cal/gm-mol or 3.7 kwhr/lb of Al. In the electrolytic process in use today, part of the fuel is provided in the form of electricity and part in the form of carbon electrodes. The reduction proceeds according to the reaction

$$Al_2O_3 + \frac{3}{2}C = Al_2 + \frac{3}{2}CO_2$$

This reaction requires a minimum of 0.33 lb of C/lb of Al or 1.37 kwhr/lb of Al of available useful work supplied by C which is oxidized to CO_2. It follows that the minimum amount of electricity is 2.33 kwhr/lb of Al or a fuel equivalent of 7.0 kwhr of fuel/lb of Al. Thus, the overall practical minimum for the electrolytic process is 8.4 kwhr of fuel/lb of Al.

In operating electrolytic aluminum plants, the specific fuel consumption is about 0.5 lb/lb of Al or 2.1 kwhr/lb of Al in the form of carbon electrodes plus an average of 7.5 kwhr/lb of Al of electricity or 22.5 kwhr of fuel/lb of Al.[29,30] Thus the total specific fuel consumption is 24.6 kwhr/lb of Al, not including the fuel required for the processing of bauxite ore and carbon electrodes. This consumption is about 3 times as large as the minimum practical requirement of 8.4 and about 6.5 times the minimum theoretical requirement of 3.7. A small amount of the difference between 24.6 and 8.4 (less than 1 kwhr/lb of Al) is not lost in the process because the products of the electrolysis are at 1800°F and some of the effluent gas is in the form of CO rather than CO_2. The relatively large difference between actual and minimum available useful works (practical or theoretical) suggests that opportunities exist for important fuel savings. These are discussed in the following sections.

5.4.2 Potential Process Improvements

5.4.2.1 Improvements in Hall Process

Analysis of the cell voltage shows that only 1.6 to 1.8 volts, out of almost 5 volts drop across the cell, is required for the basic electrolysis process.[29,31] The remainder is necessary as a result of voltage drops (resistive losses) across various electrical resistances in the cell circuit. Because the electrolysis voltage is relatively independent of current through the cell, aluminum production is approximately proportional to the current and the parasitic resistive losses are proportional to the square of the current. If follows that the fraction of electricity effectively utilized for the electrolytic reduction of Al_2O_3 increases as the cell current is reduced. (If this reduction is too great, certain heat losses and side reactions outweigh the effect of the reduction in current density.) For example, reducing the current of a typical cell from 105,000 amps to 82,000 amps would decrease electricity consumption per ton of aluminum by 16%.[29] Although such current reduction would decrease production per cell by 22%, total production can be maintained at the desired level by installing more cells, namely at the expense of higher capital costs. In general, the optimum cell current density decreases as power costs increase. Typical optimum values are about 1.1 amps/cm^2 at 2 mils per kwhr and 0.7 amps/cm^2 at 10 mils per kwhr. At the lower current density, electrical consumption is only about 12,500 kwhr per ton of aluminum.[32]

Various loss mechanisms that are present in the electrolysis cell are also subject to improvement. Anode overpotential, for example, which accounts for a power loss of 14% to 16%, might be reduced by incorporating catalytic additives in either the carbon anodes or the cryolite bath. Further improvements might be made by reducing the resistance of the anode, bus-bars, and connectors. With present current densities these resistances result in a voltage drop of about 0.9 volt and account for 19% of the consumption of electricity. The cathodic electrical connection to the molten aluminum, which is made indirectly by burying iron bars in the bottom carbon lining of the cell, causes a drop of about 0.4 to 0.75 volt. Titanium and zirconium borides and carbide and boride mixtures, which may prove to be practical substitutes as cathode materials, can reduce this cathode voltage drop to as little as 0.2 volt.[29,33,34]

The resistance of the electrolyte, which accounts for about 40% of the power consumed, can be lowered by reducing the interpolar distance or by modifying the composition of the electrolyte for improved conductivity.[35]

Even without modifying the basic production process, it is possible to achieve some fuel savings through better management of the operation. Nippon Light Metal Company reports[36] an average improvement in current efficiency of 3% with existing cells by tighter control of depth of the molten aluminum pad, distance between anode and cathode, and fractions of cryolite and alumina in the bath.

5.4.2.2 Alternative Processes

No radical departure from the essential features of the Hall electrolytic process have yet been achieved on an industrial scale. Nevertheless, several promising processes have been proposed which could not only reduce the fuel requirements of the aluminum industry, but would also permit the use of low-grade U.S. bauxite and clay deposits in place of high-grade imported bauxite ore. Among these are the following:

- ALCOA $AlCl_3$ Electrolysis Process[37,38]
- Toth Chemical Process[39]
- Plasma Reduction Process[40,41]

A relatively large-scale plant using the new ALCOA process is under construction. Kirby, Singleton and Sullivan[38] report that the potential advantages relative to the Hall process include:

1. Ability to use low-grade ores.
2. Operating temperature of $1400°F$ instead of $1800°F$.
3. Less expensive electrolyte (NaCl-KCl rather than cryolite).
4. No anode carbon consumption.
5. Higher anode current density.
6. Higher purity of output metal.
7. Less severe pollution problems (cell operates as a closed system).

According to ALCOA,[37] savings of 30% in electricity are achieved with the chloride process. Overall fuel savings would be 45×10^6 Btu/ton from the reduction in electricity, and 14×10^6 Btu/ton from the elimination of the oxidation of carbon. This would reduce the total fuel needs for the primary aluminum process from 190×10^6 Btu/ton to 131×10^6 Btu/ton.

The Toth process is a multi-step chemical technique for reducing aluminum, with electricity required only for utilities. Carbon in the form of coke is used to reduce the aluminum oxide. Because the Toth process is in an early stage of development, no fuel figures are available. A preliminary evaluation indicates that the carbon consumption may be sufficiently high to offset most of the saving of electricity relative to the Hall Process.

Another possible approach is based upon plasma-arc reduction of Al_2O_3 or $AlCl_3$. Initial studies of this concept show that substantial fuel reduction relative to the Hall process is theoretically possible. Three basic variations in the plasma arc approach have been suggested:

1. Reduction using solid carbon:

$$Al_2O_3 + 3C \longrightarrow 2Al + 3CO$$

2. Reduction using a hydrocarbon:

$$Al_2O_3 + 3CH_4 \longrightarrow 2Al + 3CO + 6H_2$$

3.   Conversion to chloride and reduction of the chloride:

$$Al_2O_3 + 3C + 3Cl_2 \longrightarrow (AlCl_3)_2 + 3CO$$

$$2\ AlCl_3 + 3H_2 \longrightarrow 2Al + 6HCl$$

Assuming 75% plasma-heat efficiency as previously demonstrated for tungsten reduction, calculated electrical consumption is in the range of 5700 to 9500 kwhr per ton of aluminum. Although this range is well below the best values required for the Hall process, it must be pointed out that no experimental data are yet available.

5.4.3 Increased Scrap Recycling

The total aluminum used for manufacturing is the sum of primary aluminum production and secondary or recycled scrap aluminum (Figure 5-4-1). In the manufacturing of products, new scrap is produced which is assumed to be completely recycled. The products can be divided into those with a very long product life, such as aluminum used in building and construction, and those with a short life, such as containers and packaging, which could be completely recycled. At the end of the product life, part of the aluminum is recycled as represented by stream (8) and part is discarded as represented by stream (7). At the present time, stream (7) is very large and stream (8) is very small.

Because refining of waste aluminum requires only 5% of the fuel needed for the Hall process, recycling is a highly effective means of reducing fuel requirements. In 1965, recycling of scrap[42] from manufacturing waste amounted to 14% of total output, but scrap from end-products that outlived their usefulness amounted to only 4.7% (0.16×10^6 tons).

We have estimated the fuel saving from maximum recycling (stream (7) = 0). For the purposes of this estimate, the end-product use of aluminum[43] has been corrected to account for exports, and average useful lifetimes for the various product categories have been assumed as shown in Table 5-4-1.

The contribution of old scrap in a given year must be related to the amount of aluminum used in the product when manufactured, that is, at an earlier year depending on the product average life. For this estimate, a ten-year doubling time is assumed for the consumption of aluminum for all products. It is also assumed that 14.2% of the material entering manufacturing is recycled as new scrap with a negligible recycling time, and that optimum recycling will result in stream (7) approaching zero.

When the above assumptions are applied to production by the aluminum industry in 1965, the maximum amount of recycled old scrap (stream (8) in Figure 5.4.1) can be calculated. The results are shown in Table 5.4.2. It is seen from the table that recycled old scrap may be increased from 0.16×10^6 tons (the actual 1965 figure) to 1.4×10^6 tons. The recycling rate then becomes 1.14/3.39 or 33.5%, instead of 4.7% for 1965. Although it may be impossible to

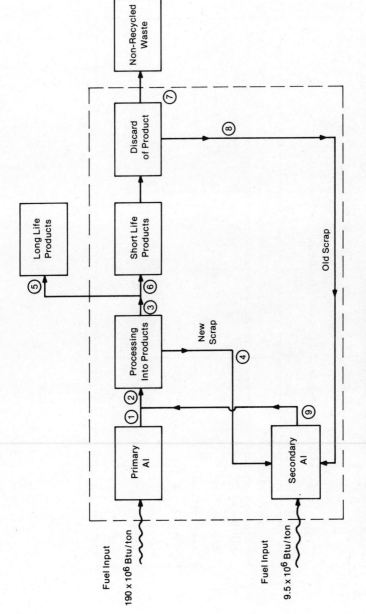

Figure 5-4-1. Simplified aluminum flow diagram.

Table 5-4-1.   Average Useful Lifetimes for Aluminum Products

End-Product Use	% of Total Consumption	Average Product Lifetime
Building and Construction	28.1%	Infinite
Other	7.4%	Infinite
Transportation	18.2%	5 years
Containers and Packaging	15.4%	2 years
Electrical	14.5%	25 years
Consumer Durables	9.8%	5 years
Machinery and Equipment	6.6%	10 years

Table 5-4-2.   Theoretical Amount of Aluminum that can be Recycled

Sector	Mass consumed in 1965 10^6 tons)	Mass available for recycling $= (M_{65})\exp(-kt)$ where $k = 0.693$ and t = average years life of products $(10^6$ tons)
Transportation	.531	0.375 (t = 5 years)
Container and Packaging	.449	0.391 (t = 2 years)
Electrical	.423	0.075 (t = 25 years)
Consumer Durables	.286	0.202 (t = 5 years)
Machinery and Equipment	.192	0.096 (t = 10 years)
TOTAL	1.88	1.14

achieve this high degree of recycling, considerable increase over the 4.7% rate of 1965 can be expected.

5.4.4 Fuel Distribution

Most aluminum reduction is carried out either in the Pacific Northwest region or in the region of the Tennessee Valley Authority where coal and hydro are the main sources of electricity. Because the extensive inter-connection of U.S. electric utilities permits the ready exchange of power between regions, aluminum production must be regarded as a load on the entire electricity grid and as a drain through that grid on scarce petroleum and gas resources. Considered in this context, the approximate flow of fuel for the aluminum industry in 1968 is shown in Figure 5-4-2. Overall demand upon

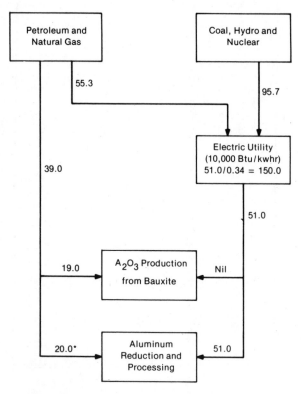

*Includes heating value of carbon electrodes.

All fuel values in 10^6 Btu per ton of aluminum.

Figure 5-4-2.   Average distribution of fuel for U.S. primary aluminum production in 1968.

petroleum and gas resources is seen to be 95.7×10^6 Btu per ton of primary aluminum, or about 50% of the total fuel consumed in aluminum production.

Total aluminum production in the U.S. in 1968 was 4.07×10^6 of which 0.15×10^6 tons (3.8%) was produced from old scrap and 0.66×10^6 tons (16.2%) from new scrap. Thus, the total fuel consumed and the average specific fuel consumption in 1968 were

$$E_{Al} = (3.26 \times 10^6)(190 \times 10^6) + (0.81 \times 10^6)(9.5 \times 10^6)$$

$$= 0.63 \times 10^{15} \text{ Btu.}$$

$$\frac{E_{Al}}{M} = \frac{0.63 \times 10^{15}}{4.07 \times 10^6} = 155 \times 10^6 \text{ Btu/ton}$$

Because total U.S. aluminum production is expected to increase to 8.9×10^6 tons by 1980, fuel demand will become 1.38×10^{15} Btu, or 119% larger than that of 1968 if present practices are not improved. Based upon a 16% recycling rate for old scrap (out of 33.5% theoretically possible), 16% recycling of new scrap (same as 1968), and a 20% reduction in fuel requirements for primary aluminum production as a result of improvements in or substitution for the Hall process, an estimate of the potential fuel needs for the aluminum industry based on known technology and for an output of 8.9×10^6 tons/yr becomes:

$$E_{Al} = (0.68)(8.9 \times 10^6)(152 \times 10^6) + (0.32)(8.9 \times 10^6)(9.5 \times 10^6)$$

$$= 0.95 \times 10^{15} \text{ Btu,}$$

and

$$\frac{E_{Al}}{M} = \frac{0.95 \times 10^{15}}{8.9 \times 10^6} = 106 \times 10^6 \text{ Btu/ton.}$$

5.5 COPPER

Unlike aluminum ore, copper minerals are easily reduced with relatively little consumption of fuel. On the other hand, copper ores range at present from about 0.7 to 2% copper by weight as compared with bauxite ores for aluminum which are generally about 26% aluminum. The depletion of high assay ores over the past 30 years is a major cause of the increase in the cost of copper. In terms of constant dollars, copper prices have risen by a factor of more than two since 1940, whereas aluminum prices in the same period have actually declined by almost one half.[44]

Approximately 15 copper minerals (of the 164 known) are commonly found in ore deposits. The ores containing these minerals can be classified as sulfides and oxides, with most copper now obtained from the sulfide ores. Some ores also contain other valuable metals which may be recovered along with the copper. Two general methods are used at present for the production of copper metal:

- Pyrometallurgical or smelting operation, followed by electro-refining.
- Hydrometallurgical operation in which the ore is first treated with a suitable solvent which selectively dissolves the copper, followed by either precipitation of the copper by metallic iron or electrowinning of the copper from the solution.

The choice of process depends on, among many factors, the type of ore and its copper content. Smelting, which is commonly used with the sulfide ores, is currently by far the more important of the two methods in that it accounts for about 90% of world production. The hydrometallurgical process has been applied primarily to oxide ores in which the copper is readily soluble in sulfuric acid and to tailings from previous mining and smelting operations.

5.5.1 Smelting or Pyrometallurgical Process[45-55]

High-assay sulfide ore (0.7–2% Cu) is concentrated by flotation to yield a feedstock which is 15% to 35% copper. This concentrate is then roasted in multiple-hearth furnaces in which oxidizing reactions take place. The ore is heated in the presence of air to a temperature of about 1000°F, with little or no fusion taking place, in order to eliminate sulfur and other impurities. Because the reactions are highly exothermic, high-sulfide ores can be roasted without externally supplied fuel.

One method for reducing the fuel requirements in roasting is used in the Chambishi process[48] where two streams of ore concentrate, one with insufficient sulfur for autogenous roasting and the other with excess sulfur, are blended to provide an input to the roasting furnace with the proper sulfur content. Another method, which is used with high sulfur ores, is to combine the roasting and smelting operations so that the energy that becomes available by exothermic reactions during roasting is used directly to replace consumption of fuel in the smelting operation.[54] The best roasting temperature is in the range of 1290°F–1380°F, and the calcines produced are charged hot to the reverberatory smelting furnace to reduce its fuel requirements.

In the smelting operation, a molten sulfide of copper and iron, or matte, is produced. The smelting operation takes advantage of the high affinity of copper for sulfur. Thus, the sulfur combines preferentially with copper, rather than iron, forming cuprous sulfide, and the oxygen combines preferentially with iron, forming ferrous oxide. The cuprous and ferrous sulfides dissolve in each other, with a reduction in the melting point relative to the pure sulfides (melting point about 1800°F), forming the fused matte with a higher specific gravity than the accompanying waste slag. The smelting is carried out in a fuel-fired reverberatory furnace. Combustion air is preheated to 700°F by a regenerative air preheater. In addition, a waste heat boiler is generally used to recover energy from the exhaust gases, which then leave the waste heat boiler at a temperature of about 650°F. The molten matte contains 25%–45% copper, depending on the assay of the input ore.

The smelting operation at about 2400°F has by far the highest fuel consumption of any part of the process. The fuel requirement is generally given in terms of Btu per ton of solid charge. It is strongly dependent upon the copper content of the charge and whether the reverberatory furnace is charged with hot calcine or wet concentrate not previously roasted. Average values for the smelting operation are 32×10^6 Btu and 375 kwhr of electricity per ton of copper metal. Total equivalent fuel for smelting is therefore 36×10^6 Btu/ton.

The matte from the smelting operation goes to the converter for the production of blister copper which is about 98% copper. Silica containing flux is added to unite with the iron of the matte and form a suitable slag. In the converter, air is forced through the molten matte. Almost all of the energy required in the conversion is derived from oxidation of the ferrous sulfide. The blister copper from the converters is further fire-refined to remove impurities and, by reducing the sulfur and oxygen content, to prevent excessive degassing during casting. The fuel requirement for the fire-refining, which is the final refining operation for about 14% of the world's copper, is about 5×10^6 Btu/ton. The other 86% is further refined electrolytically.

Electrolytic refining is used for production of very pure copper and for recovery of any precious metal impurities in the copper.[56] Electrical work for the electrodeposition process, including auxiliary equipment, is about 280 kwhr per ton of copper, or 2.8×10^6 Btu/ton in equivalent fuel. The resultant copper cathodes are then further refined by melting at a fuel expense of 3.3×10^6 Btu/ton to achieve a purity of 99.9%.

Total fuel consumption for copper production by pyrometallurgical methods is thus 47.1×10^6 Btu/ton.

5.5.2 Leaching or Hydrometallurgical Process[57-59]

An increasing amount of primary copper production is being recovered from low-grade copper-bearing ores and waste rock that cannot be processed economically by conventional smelting. All the processes used for these materials employ hydrometallurgical methods, which dissolve or leach the copper from the ore, followed by separation of the metal from the solutes. Because of the predominance of low-grade materials, the trend for the future appears to be toward hydrometallurgical processes. Major steps in the process are leaching, purification, concentration, and recovery. Several alternative techniques are used for each of these three steps. Those of the first step include (a) leaching in place, (b) heap leaching, and (c) vat leaching. If these procedures are fully developed, the estimated 75 million short tons of copper contained in U.S. ores averaging 0.86% copper would be supplemented by another 58 million short tons of copper in ore deposits assaying less than 0.47%.

The amount of purification and concentration required depends on the copper and impurity content of the leach solution and on the procedure used for recovery of copper metal from the solution. A relatively new process for preparation of the electrolyte for the electrowinning of copper is solvent extraction or liquid-liquid ion exchange.[58,59] In essence, the solvent extraction plant is a chemical refinery. All of the material handled, except for the finished product, is in liquid form. The purpose of the process is to reduce the iron content of the electrolyte which goes to the electrowinning plant, and thereby reduce the electrical power required.

In the conventional process for recovery of copper metal from dilute leach solutions, which is known as cementation, the copper-bearing acid solution is brought into contact with either scrap iron or specially-prepared sponge iron. The iron dissolves and displaces the copper, which precipitates out. Theoretically, 1 pound of copper should be precipitated by 0.878 pound of iron. In practice, 1.3–4.0 pounds of iron are required per pound of copper because of the dissolution of iron by sulfuric acid and ferric sulfate. The cement copper produced, being only 80 to 90% copper, requires additional processing.

Another recovery process involves hydrogen reduction of the $CuSO_4$ acid solution. This is carried out in an autoclave at a hydrogen partial pressure of 350 psi and temperature of 250–285°F. The copper is precipitated as a high-purity metallic powder and the sulfuric acid required for leaching is regenerated. The demand for copper powder, which sells for about 15 cents per pound more than the metal, is increasing.

The required fuel for the hydrometallurgical process is relatively independent of the copper content of the ore. Typical plants operate at 2600 to 3200 kwhr per ton. The average fuel demand is about 29×10^6 Btu/ton.

In 1968 U.S. copper production amounted to 3.1×10^6 tons, of which 43% was supplied by scrap which requires only about 4×10^6 Btu/ton for remelting. Total fuel consumed by the industry in smelting 90% of the primary copper and electrorefining 86% was 0.08×10^{15} Btu, or less than 0.4% of U.S. industrial fuel.

In considering improvements in the use of fuel for copper production, allowance must be made for the low copper content of current ore deposits and the decrease in copper content year by year as the deposits of higher grade are depleted. This decrease in the quality of ore along with an increase in the cost of fuel will result in greater emphasis upon hydrometallurgical processes.

Other steps which should be taken to reduce fuel consumption include the following:

1. Further development of continuous processes for roasting, smelting, and converting, so that maximum use is made of the inherently exothermic reactions that occur.
2. Oxygen enrichment of the air feed for smelting to eliminate fuel requirements in processing high-sulfur ores.
3. Greater use of boilers for recovery of waste heat.
4. Reduced current density in electrorefining operations at the expense of higher capital cost.
5. Reduced current density in electrowinning of copper from leach solution.
6. Solvent extraction to transfer copper selectively from leach solution to electrolyte, thereby reducing impurities which adversely affect fuel consumption in the electrowinning operation.

Because of its high scrap value, copper is already so extensively recycled that little improvement in the fuel consumption for copper production can be expected from that quarter. Nevertheless, despite a gradual deterioration in the assay of available copper ores, implementation of the improvements listed above could reduce fuel requirements by at least 30%.

5.6 CEMENT

5.6.1 Process Description and Fuel Consumption

In the manufacture of cement, the finely ground raw materials (limestone, sand and clay) are heated in a long, rotating, tubular kiln to decarbonize the limestone and to form a glassy cement clinker. Portland cement is made by mixing a small amount of gypsum with finely ground clinker.

A typical modern plant produces 750,000 tons of cement per year. Raw materials required to produce one ton of cement clinker are as follows:[60]

Limestone ($CaCO_3$)	1.2
Clay ($Al_2O_3 \cdot 2SiO_2 \cdot 2H_2O$)	0.3
Sand (SiO_2)	0.1
	1.6 tons

The limestone is finely ground and blended with sand and clay. Blended raw materials are then fed to a firebrick-lined cylindrical kiln about 16 feet in diameter and 400 feet in length. The kiln, which is rotated at 1 rpm, has a slope so that raw materials flow down the length of the kiln in a tumbling motion counter to the flow of products of combustion from the burner. As they travel down the kiln, the materials are preheated and dried by products of combustion. As the mix reaches 1200°F, the limestone decomposes into carbon dioxide (CO_2) and calcium oxide (CaO). Upon further heating to 2300°F, dicalcium and tricalcium silicates ($2CaO \cdot SiO_2$ and $3CaO \cdot SiO_2$) and tricalcium aluminate ($3CaO \cdot Al_2O_3$) are formed.[61]

Rotation of the kiln causes the resulting glassy phase to appear as small roundish balls of cement clinker. These fall into an air blast which optimizes cement properties and recovers waste heat from the clinker. Cooling air from the clinker is supplied as pre-heated secondary air to the kiln burner. The cooled clinker is ground to 325 mesh, and is blended with a small percentage of gypsum which retards the setting time for the concrete mix.

In the cement-making process 80% of the total fuel consumed provides heat while the remainder produces electricity which is used primarily for grinding incoming ore and finished clinker. Heat is provided by coal, oil, or gas in kilns, approximately 40% of which are designed to burn any of these fuels. In multi-fuel installations, gas is usually the primary fuel. Fuel requirements range from about 5.8×10^6 Btu/ton of cement in plants with gas firing to 7.3×10^6 Btu/ton of cement in plants with coal firing. In 1970 coal provided 53% of the total fuel supplied to U.S. cement kilns and preheaters, gas provided 46% of the total, and oil the remainder.[62]

The approximate distribution of fuel for the industry[63] in 1968 is shown in Figure 5-6-1. Total fuel input is seen to be 7.9×10^6 Btu/ton. For the same period, total U.S. cement production was 72×10^6 tons. Thus, fuel consumption for cement making in 1968 amounts to 0.57×10^5 Btu, or about 2.5% of all U.S. industrial fuel. Approximately 44% of this fuel comes from the more scarce petroleum and natural gas resources, a fraction somewhat better than the 60% fraction for U.S. industry in general. The Portland Cement Association[64] expects cement production to increase to 92×10^6 tons by 1980.

5.6.2 Analysis of Available Useful Work

An estimate of the available useful work for the formation of Portland cement can be obtained as follows. Limestone and clay can be considered as the basic calcareous and argilaceous materials, respectively. The major component of limestone is calcium carbonate ($CaCO_3$). On the other hand, clays are composed of a number of different hydrated aluminosilicates with ratios SiO_2/Al_2O_3 ranging from 2 : 1 to 5 : 1; some clays also contain ferric oxide as an essential constituent. For present purposes, attention may be restricted to kaolinite ($2SiO_2 \cdot Al_2O_3 \cdot 2H_2O$), one of the simplest clay compounds.

Analysis of typical cement yields the composition[65] shown in Table 5-6-1. It has been shown,[65,66] however, that this composition can be represented in terms of the primary compounds $2CaO \cdot SiO_2$, $3CaO \cdot SiO_2$, $3CaO \cdot Al_2O_3$, $4CaO \cdot Al_2O_3Fe_2O_3$ and CaO. Neglecting the contributions of

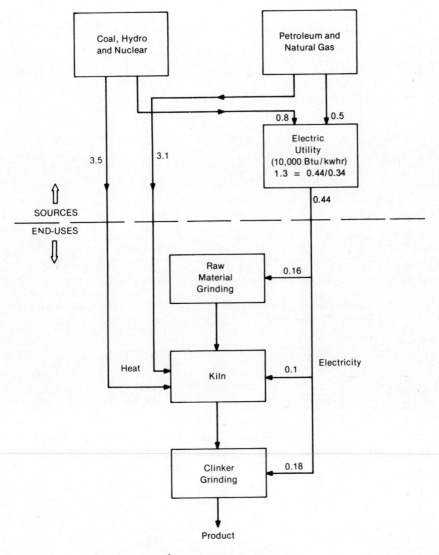

All fuel values in 10^6 Btu per ton of cement.

Figure 5-6-1. Sources and applications of fuel in U.S. cement industry in 1968.

Table 5-6-1.   Composition of Typical Portland Cement

	Percent
CaO	64.10
SiO_2	22.90
Al_2O_3	4.50
Fe_2O_3	3.11
MgO	0.79
TiO_2	0.24
Na_2O	0.54
K_2O	0.64
SO_3	2.37
Unaccounted	0.81
	100.00

Table 5-6-2.   Composition of Portland Cement in Terms of Primary Compounds

Primary Compound	Percent
$2\,CaO \cdot SiO_2$	36.5
$3\,CaO \cdot SiO_2$	45.1
$3\,CaO \cdot Al_2O_3$	7.2
$4\,CaO \cdot Fe_2O_3$	10.2
CaO	1.0
	100.0

magnesia, alkalis, and titania, the composition can be normalized[66] for the primary compounds as shown in Table 5-6-2.

Thus, the material balance for the overall reaction for the formation of 10^3 kg of clinker will be as shown in Figure 5-6-2. For this reaction, with

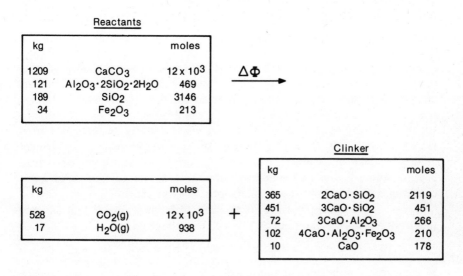

Figure 5-6-2.   Overall reaction for portland cement formation.

reactants and products at room temperature, the change $\Delta\Phi$ in available useful work is found to be:

$$\Delta\Phi = 0.8 \times 10^6 \text{ Btu/ton of clinker.}$$

5.6.3 Recommendations for Fuel Savings

The cement industry has gradually been improving the effectiveness of its use of fuel. For example, average fuel consumption in the kiln has declined from 6.9×10^6 Btu/ton in 1967 to 6.6×10^6 Btu/ton in 1972, the greatest improvements occurring in gas-fired units. Despite this trend, the specific fuel consumption in the U.S. cement industry is still two and a half times higher than the best figure achieved in European and Japanese plants, and about an order of magnitude larger than the minimum required for the process. Several factors account for this difference:

1. U.S. industry continues to produce 60% of its cement using the wet process for grinding and blending of feedstock; this process adds 30% of water to the raw materials. About 1.0×10^0 Btu/ton ot cement is therefore wasted in evaporation of water.
2. In most U.S. kilns burners are not located in the region of highest endothermic reactions.
3. Convection and radiation losses from U.S. kilns account for about 17% of total fuel input, an unnecessarily high fraction.
4. U. S. plants recover less waste heat in general and in particular recycle less heat from exhaust gas and clinker than foreign plants.

The technology for efficient fuel management in the cement industry is readily available, primarily from overseas equipment manufacturers. The reluctance to use it in the U.S. comes from the relatively high capital investiment required (about $40 million for a modern cement plant producing 0.75 million tons of cement) and the relatively small effect of present fuel costs upon cement price. If U.S. industry were to adopt the most modern foreign practice,[67,68] average fuel input to the kiln could be reduced from 6.6×10^6 to about 3.6×10^6 Btu/ton of cement, and electric utility requirements from 1.3×10^6 to about 0.9×10^6 Btu/ton of cement. Moreover, the cement industry, unlike most others, can use high-sulfur coal as the preferred fuel, because sulfur in the fuel combines with the raw mix to form solid sulfur compounds which are to a degree tolerable in finished cement.

If advantage is taken of available technology, the cement industry fuel requirements for an output of 92×10^6 tons/yr would be

$$92 \times 10^6 \text{ tons/year} \times 4.5 \times 10^6 \text{ Btu/ton} = 0.4 \times 10^{15} \text{ Btu/year.}$$

In comparison with 1968, it follows that the cement industry could increase production by 28%, while reducing fuel needs by 30%.

REFERENCES

1. Harold E. McGannon, "The Making, Shaping and Treating of Steel," U.S. Steel, Ninth Edition (1971).

2. *Annual Statistical Report*, American Iron and Steel Institute, Washington, D.C. (1972).

3. T. J. Ess, "The Modern Coke Plant," Iron and Steel Engineer, 25(1), pp. C3-C37 (January 1948).

4. G. S. Lucenti, "The Continuous Casting of Beam Blanks at the Algomen Steel Corp., Ltd.", Iron and Steel Engineering, 46(7), pp. 83-100 (July 1969).

5. E. H. Gott, "The Economic Importance of Continuous Casting of Steel Slabs," Proceedings of the Third Annual Conference of the International Iron and Steel Institute, pp. 102-115 (1969).

6. W. Trinks and M. H. Mawhinney, *Industrial Furnaces*, 5th Edition, John Wiley, New York (1961).

7. From Japan Iron and Steel Federation Statistics.

8. Takao Yamada, "Blast Furnace Practices at Mizushima Works," Iron and Steel Engineer, p. 53 (November 1970).

9. Ailleen Cantrell, "Annual Refining Survey," Oil and Gas Journal, pp. 99-125 (April 2, 1973).

10. W. L. Nelson, "What is Refining Complexity," Oil and Gas Journal, p. 93 (March 7, 1973).

11. FPC Docket CP-72-6, Table S-206.

12. M. G. Whitcomb, Jr. and F. M. Orr, "Plan Plant Energy Conservation," Hydrocarbon Processing, pp. 65-66 (July 1973).

13. J. E. Hayden and W. H. Levers, "How to Conserve Energy While

Building, Expanding a Refinery," The Oil and Gas Journal, pp. 109-116 (May 21, 1973).

14. *The Pulp and Paper Industry*, Patterns of Energy Consumption in the United States, Stanford Research Institute, January 1972.

15. W. F. Morse, *Process Energy Requirements in the Pulp and Paper Industry*, American Gas Association, Inc. (1967).

16. *Pulp, Paper, and Board*, U.S. Dept. of Commerce, Vol. XXIX, No. 1 (April 1973).

17. R. J. Slinn, *Sources and Utilization of Energy in the U.S. Pulp and Paper Industry*, American Paper Institute (March 1973).

18. L. D. Mourer and A. W. Peterson, *Power and Consumption Values for Pulp and Paper Mills*, Tappi, Vol. 49, No. 3 (March 1966).

19. *Boiler and Recovery Units for Pulp and Paper Industry*, Combustion Engineering, Reference Book, Chapter 27, G. R. Fryling.

20. *Chemical and Heat Recovery in the Paper Industry*, Steam, Its Generation and Use. Chapter 20, Babcock and Wilcox Company.

21. L. Elmernus, The Application of By-Product Power Rate, Tappi, Vol. 55, No. 5 (May 1972).

22. *Energy Conservation in the Pulp and Paper Industry*, Swedish Pulp and Paper Mission to North America (November 1973).

23. O. J. Kalmes, *A Review of the Lodding Paper Making Project*, M/K Systems, January 1973.

24. J. N. Stephenson, "Pulp and Paper Manufacture, Preparation and Treatment of Wood Pulp," Vol. I, McGraw-Hill, New York (1950).

25. Perry's "Handbook of Chemical Engineering," 4th Edition, McGraw-Hill, New York.

26. "Fire Prevention Handbook," 13th Edition, National Fire Protection Association, (1969).

27. G. L. Lanz, "Thermodynamic Properties of Organic Compounds," Academic Press, New York, (1967).

28. G. Odian, "Principles of Polymerization," McGraw-Hill, New York, (1970).

29. M. Van Lancker, Metallurgy of Aluminum Alloys, Chapter II—The Industrial Production of Alimnum, Chapman and Hall, Ltd., London (1967).

30. E. A. Hollingshead and V. A. Braunworth, "Laboratory Investigation of Carbon Anode Consumption in the Electrolytic Production of Aluminum," pages 31-50; G. Gerard, Editor, Extractive Metallurgy of Aluminum, Vol. 2, Aluminum, Interscience Publishers, New York (1963).

31. D. Gilroy, "The Electrowinning of Metals," Chapter 6, A. Kuhn, Editor, Industrial Electrochemical Processes. Elsevier Publishing Company, Amsterdam, The Netherlands (1971).

32. R. Scalliet, "Considerations on Modern Cells with Prebaked Anodes," Extractive Metallurgy of Aluminum, Vol. 2. Aluminum, G. Gerard, Editor, Interscience Publishers, New York (1963).

33. C. E. Ransley, "The Applications of the Refractory Carbides and Borides to Aluminum Reduction Cells," Extractive Metallurgy of Aluminum, Volume 2. Aluminum, G. Gerard, Editor, Interscience Publishers, New York (1963).

34. J. P. Givry, "Technical and Economic Aspects of Aluminum Cell Conductors," Extractive Metallurgy of Aluminum, Vol. 2, Aluminum, G. Gerard, Editor, Interscience Publishers, New York (1963).

35. R. A. Lewis, Chemical Engineering Progress, *56*, 78 (1960).

36. T. Bessho, and H. Suetake, "Appraisal of the Operation of Aluminum Reduction Cells for Maintaining High Current Efficiency," pp. 172-188, R. M. Kibby, Editor, Proceedings of the Extractive Metallurgy Division Symposium, Chicago, Illinois (December 1967).

37. *Chemical Engineering*, p. 61, June 11, 1973.

38. D. E. Kirby, E. L. Singleton and T. A. Sullivan, "Electrowinning of Aluminum from Aluminum Chloride," Bureau of Mines, Report of Investigations 7353, U.S. Department of the Interior (1970).

39. C. Toth and A. Lippman, "The Quest for Aluminum," *Mechanical Engineering*, pp. 24-28 (September 1973).

40. R. K. Rains and R. H. Kadlec, "The Reduction of Al_2O_3 to Aluminum in a Plasma," Metallurgical Transactions, *1*(6), pp. 1501-1506, (June 1970).

41. Personal Communication, Robert D. Davis, President, Plasmachem, Santa Anna, California.

42. Statistical Abstracts of the United States, 88th Annual Edition, 1967, Superintendent of Documents, U.S. Government Printing Office, Washington, D.C.

43. "The World Almanac," 1973 Edition, Newspaper Enterprise Association, New York, New York.

44. The U.S. Energy Problem: Vol. II, Appendices—Part B by Inter Technology Corporation (November 1971), Distributed by National Technical Information Service, 5285 Port Royal Road, Springfield, Va. 22151, Accession No. PB-20719.

45. J. N. Anderson, "Reverberatory Furnace and Convertor Practice at the Noranda and Gaspe Smelters," presented in *Extractive Metallurgy of Copper, Nickel, and Cobalt*, Edited by P. Queneau, Interscience Publishers, New York (1961).

46. M. G. Flowler, "Smelting Practices of Phelps Dodge in Arizona," presented in *Extractive Metallurgy of Copper, Nickel and Cobalt*,

47. "Patterns of Energy Consumption in the United States," by Stanford Research Institute, Menlo Park, California; Stock Number 4106-0034, Superintendent of Documents, U.S. Government Printing Office, Washington, D.C. 20402.

48. L. R. Verney, J. E. Harper and P. N. Vernon, "Development and Operation of the Chambishi Process for the Roasting, Leaching, and Electrowinning of Copper," pp. 272-305, Extractive Metallurgy Division Symposium on Electrometallurgy, Cleveland (1968), Met. Soc. AIME, New York.

49. W. B. Boggs, "Roasting, Smelting, and Converting," Ch. 4, *Copper—The Science and Technology of the Metal, Its Alloys, and Compounds*, A. Butts, Editor, Reinhold Publishing Corporation, New York (1954).

50. G. Bridgestock, "How to Limit SO_2 Emissions with the Flash Smelting Process," Eng. and Min. J., Vol. 172, No. 4, pp. 120-122, (April 1971).

51. Staff, The International Nickel Co., "The Oxygen Flash Smelting Process of the International Nickel Co." Trans. Can. Inst. Mining Met., Volume 58, p. 158 (1955), Canadian Mining Metallurgical Bulletin, Volume 48, p. 292 (1955).

52. N. J. Themelis, G. C. McKerrow, P. Tarassoff, and G. D. Hallett, "The Noranda Process," Journal of Metals, pp. 25-32 (April 1972).

53. B. Stevens, "Continuous Copper Smelting," Chemical and Process Engineering, pp. 28-30 (January 1972).

54. R. B. Worthington, "Autogenous Smelting of Copper Sulfide Concentrate," Report of Investigations 7705, U.S. Dept. of the Interior, Bureau of Mines, Washington, D.C. (1973).

55. O. Barth, "Electric Smelting of Sulfide Ores," Extractive Metallurgy of Copper, Nickel, and Cobalt, P. Queneau, Editor, Interscience Publishers, New York (1961).

56. C. W. Eichrodt and J. H. Schloen, *Electrolytic Copper Refining*, Chapter 8, "Copper—The Science and Technology of the Metal, Its Alloys, and Compounds," A. Butts, Editor, Reinhold Publishing Corporation, New York (1954).

57. A. Miller, "Process for the Recovery of Copper from Oxide Copper-Bearing Ores by Leach, Liquid Ion Exchange, and Electrowinning at Ranchers Bluebird Mine, Miami, Arizona, pp. 337-367 of *The Design of Metal Producing Processes*, R. M. Kibby, Editor, American Institute of Mining, Metallurgical, and Petroleum Engineers, Inc., New York (1969).

58. R. R. Nelson and R. L. Brown, "The Duval Corporation Copper Leach, Liquid Ion Exchange Pilot Plant," pp. 324-336 of *The Design of Metal Producing Processes*, R. M. Kibby, Editor, American Institute of Mining, Metallurgical, and Petroleum Engineers, Inc., New York (1969).

59. A. J. Monhemius, "Trends in Copper Hydrometallurgy," *Chemical and Process Engineering,* pp. 65-68 (January 1970).

60. Gygi, H., "The Thermal Efficiency of the Rotary Cement Kiln," *Cement and Lime Manufacture*, Vol. 10, No. 11, p. 295 (Nov. 1937).

61. F. M. Lea, *The Chemistry of Portland Cement and Concrete*, pp. 117, London (1956).

62. B. C. Brown, *Cement*, Bureau of Mines Minerals Yearbook (1971).

63. B. C. Brown, *Cement,* Bureau of Mines Minerals Yearbook (1970).

64. R. D. MacLean, "Portland Cement, A Changing Industry," Roch Products, Vol. 76, No. 1, p. 93 (January 1973).

65. F. M. Lea, "Chemistry of Cement and Concrete," Edward Arnold Publishers, London, (1956).

66. R. H. Bogue, Industrial and Engineering Chemistry, Analytical Edition, *1*, (4), 192, (1929).

67. R. Semler, Jahresbericht der Deutchen Zementindustrie, e. V. (1973).

68. P. Weber, "Heat Transfer in Rotary Kilns," Bauverlag GmbH Wiesbeden, Berlin (1963).

About the Authors

Elias P. Gyftopoulos heads the Nuclear Engineering Department at the Massachusetts Institute of Technology, where he has been involved in teaching and research for almost twenty years. Professor Gyftopoulos received the Sc. D. from MIT, and has published over eighty journal articles in such areas as reactor dynamics, surface physics and plasma physics.

Lazros J. Lazaridis is manager of Thermo Electron Corporation's Thermal Systems Group, which is primarily concerned with the augmentation of heat transfer through the application of advanced technologies. Mr. Lazaridis has been granted numerous patents for his developments in heat transfer processes and products.

Thomas F. Widmer, who holds degrees in both mechanical engineering and automotive engineering, is Vice President, Engineering, at Thermo Electron Corporation. Mr. Widmer's previous experience was with the Westinghouse Atomic Power division, and with the Missile and Space Division of General Electric.